MIMESIS
INTERNATIONAL

ATMOSPHERIC SPACES
n. 16

ATMOSPHERIC SPACES

What is an "Atmosphere"?

According to an aesthetic, phenomenological and ontological view, such a notion can be understood as a sensorial and affective quality widespread in space. It is the particular tone that determines the way one experiences her surroundings.

Air, ambiance, aura, climate, environment, genius loci, milieu, mood, numinous, lived space, Stimmung, but also Umwelt, ki, aida, Zwischen, in-between – all these words are names hiding, in fact, the founding idea of atmospheres: a vague ens or power, without visible and discrete boundaries, which we find around us and, resonating in our lived body, even involves us.

Studying atmospheres means, thus, a parte subjecti, to analyse (above all) the range of unintentional or involuntary experiences and, in particular, those experiences which emotionally "tonalise" our everyday life. A parte objecti, it means however to learn how atmospheres are intentionally (e.g. artistically, politically, socially, etc.) produced and how we can critically evaluate them, thus avoiding being easily manipulated by such feelings.

Atmospheric Spaces is a new book series whose aim is to become a point of reference for a community that works together on this philosophical and transdisciplinary subject and for all those whose research, more broadly, is involved in the so-called "affective turn" of the Social Sciences and Humanities.

Robert Josef Kozljanič

GENIUS LOCI

On Neophenomenology and Hermeneutics of a Multi-Layered Appearance

MIMESIS
INTERNATIONAL

The book is published with the support of the "Stiftung Neue Phänomenologie" (www.snp-online.net) and the "Gesellschaft für Neue Phänomenologie" (www.gnp-online.de)

Translated by Richard Falzmann

www.mimesisinternational.com
e-mail: info@mimesisinternational.com

Book series: *Atmospheric Spaces,* n. 16

Isbn: 9788869775116

P.I. C.F. 0241937030

Cover image: Nymph cave near Vari, Attica, by R. J. Kozljanič.

CONTENTS

1. Introduction and Overview 9

2. Outline of a Cultural History and Phenomenology of the Genius Loci 15

3. Genius-Loci-Related Urban Development – By Reference to Ancient Rome 57

4. The Felt-Body-Phenomenological Concept of "Spatially Effused Atmospheres" of Hermann Schmitz 81

5. Walter Benjamin's Concept of Aura and the Aura of a Place – By Reference to Judengasse in Frankfurt 105

6. Current Phenomenological Debate: Schmitz, Seamon, Kozljanič, Griffero 129

7. Problems of Approach and Methodology: Environmental (Neo)phenomenology, Thick Descriptions and Deep Hermeneutics 135

8. Aesthetic Revision of Hermann Schmitz's Neophenomenological Method 141

9. Pathic Dialogue with Nature: Resonating, Corresponding, Poeticising. Pathic Aesthetics in Bettina von Arnim and Ludwig Klages 151

Picture Credits 163

Bibliography 165

1.
Introduction and Overview

1.1 *"Much romantic ado about nothing?"*

For more than 30 years now, I have been studying the phenomenon of genius loci, both theoretically and practically. Some accuse me of having a somewhat exaggerated penchant for romanticism. This accusation is more or less true. But that's not a bad thing. Quite the contrary. I am happy to face this accusation, especially in anti-romantic times like these. To provide a counterbalance. Against the oppressive economisation, digitalisation and quantification of our days. And, of course, against a pseudo-romantic, capitalist aestheticisation of almost everything; an over-aestheticisation that has already become second nature to many of us (cf. Böhme 2017b; and: Borch 2014, 7-16).

Paul Klee once said, "In this world, I am not tangible at all. For I dwell just as much with the dead as with the unborn. A little closer to the heart of creation than usual. And still not close enough." For me, it would have to be: "In the prosaic realm, I am not tangible at all. For I dwell just as much with poetic people as with nature-loving people. A little closer to the heart of creation than usual. And still not close enough."

Nevertheless, in recent years I have also dealt with very prosaic topics: for example, power politics and criticism of power (cf. Kozljanič 2022a). Criticism of power and romanticism are often mutually exclusive. But this does not have to be the case (cf. Kozljanič 2022b, 2024c). For in my opinion, the romantic notion of a more beautiful, more natural and at the same time more humane world – whether this lies in the past, present or future – is indispensable for a positive alternative. Criticism of power that merely remains stuck in the defence of what is to be rejected is not enough. For we should not only know what we reject and deny, we should at least also have an idea of what we wholeheartedly desire and affirm. And this includes – as I was able to show in the text "Living, dwelling, feeling. On the homeland-making function of 'heart-warming' places" (Kozljanič 2015b) – places with characteristic atmospheres and expressive physiognomies, places with a genius loci.

As I wrote in my correspondence with Rudolf Gaßenhuber: "It is quite obvious: We cannot completely escape the time and the zeitgeist we were born in. And yet: As long as humans – especially those who are civilisationally thrown off (nature) track and most of all (post) modern humans – are and remain masters of the simultaneity of the nonsimultaneity there is hope that some time this simultaneity of the nonsimultaneity does not have, as it often is the case, the effect of producing an alienation of nature but friendship with it; as a utopian past that breaks into the present or is longed for. What is essential at last is a viable, down to earth, romantic utopian dream. Thus we could modify Ernst Bloch's remark: 'which seems to everyone in childhood and in which no one has ever been: homeland'[1] into 'what seems to everyone in childhood, what some people see as a romantic premonition on the horizon, but where no Westerner has been since the Axial Age: homeland" (Gaßenhuber/Kozljanič 2024b, 55).

In 2004, I published my two-volume monograph "Der Geist eines Ortes. Kulturgeschichte und Phänomenologie des Genius Loci" ("The Spirit of a Place. Cultural History and Phenomenology of the Genius Loci.") The fact that there is now a more reception-friendly English abridged version is thanks to Tonino Griffero. At the same time, this book is more than just an abridged version of my dissertation: it represents, so to speak, a more mature and, in some respects, more concise summary – with new insights and perspectives.

1.2 *Overview of chapters 2-9*

The *second chapter* outlines the cultural history of genius loci from the ancient Romans and Greeks to the present day. The fact that I am concerned not only with the history of the term, but even more so with the history of the phenomenon, is indicated by the heading "Outline of a cultural history and phenomenology of the genius loci". Strictly speaking, the 'phenomenological' aspect lies in the fact that I repeatedly ask about the everyday experiences that people of different historical eras had with 'their local spirits' and that I take these experiences seriously. In addition to the phenomenological advantage, this also has a practical advantage: it is in and through these experiences that the respective relationships between the

1 In German: "das allen in der Kindheit scheint und worin noch niemand war: Heimat". Bloch, E. (1973). Das Prinzip Hoffnung, vol 3, Frankfurt a. M.: Suhrkamp, p. 1628.

people of the respective eras and the places that are relevant and significant to them become tangible. This then allows us to draft a history of human-place relationships.

In the *third chapter*, "Genius loci-related urban development – by reference to ancient Rome," I turn my attention to genius loci sites that are dominated and created by humans. Whereas in the more natural genius loci sites, what had become natural (the "geological" and "climatic", as well as flora and fauna, of course) played a decisive role in shaping them, in the artificial sites it is what has become historical (e.g. buildings) and culturally remembered (whether in the form of origin myths or fateful historical events). Both should be taken into account, because otherwise the planning intervention runs the risk of "exorcising" the spirit of a place.

The *fourth chapter* is devoted to Hermann Schmitz and his New Phenomenology. The focus is on scenic and local 'atmospheres' as found in genius loci sites. This chapter is entitled "The felt-body-phenomenological concept of 'spatially effused atmospheres' of Hermann Schmitz". In this chapter, I attempt to show that and to what extent the neophenomenological concept of atmosphere allows us to talk about landscapes and places in a much more differentiated way than was previously possible.

"Walter Benjamin's concept of aura and the aura of a place – by reference to Judengasse in Frankfurt" – this is the title of the *fifth chapter*. Benjamin's concept of aura offers the possibility, even in post-mythical and secular times, to speak of a quasi-numinous charisma that some genius loci sites still possess today. In examining Benjamin and the reception of his concept of aura by Gernot Böhme and Michael Hauskeller, I present the five most important components of Benjamin's concept of aura: first the 'grown' uniqueness of the object, second the sensual presence of the original, third the aesthetically-devout-contemplative distance of the beholder, fourth the singular-auratic emotional agitation of the beholder, and fifth the multi-layered-remembering testimony of an object.

In the *sixth chapter*, "Current phenomenological debate: Schmitz, Seamon, Kozljanič, Griffero," I take a brief look at the latest developments in the field of genius loci phenomenology. Once again, it is Hermann Schmitz – this time not only with his concept of atmosphere, but also with his concept of situation – who has contributed decisive and illuminating terms. Terms that – directly or indirectly – many of the more recent efforts build on. In addition to David Seamon's multi-perspective approach, Tonino Griffero's "affordance" concept of "pathic aesthetics" seems particularly promising to me. I will return to this in the ninth chapter.

It is not until the *seventh* and *eighth chapters* that I turn to methodological questions ("7. Problems of approach and methodology: Environmental (neo)phenomenology, thick descriptions and deep hermeneutics" – "8. Aesthetic revision of Hermann Schmitz's neophenomenological method"). Why not earlier? Why not right at the beginning? This has to do with my hermeneutic spiral approach, a kind of spiral-like approach to the phenomena being sought: 1. Review and outline the various terms relevant to genius loci. 2. However, these terms function as preliminary terms until they have been confronted with the respective genius loci experiences, clarified, and differentiated. 3. From these terms and experiences emerge – hermeneutically, i.e., understanding and clarifying meaning – the various genius loci concepts. 4. From these clarified and established genius loci concepts, it is possible to read and evaluate – in a cultural-historical manner – the relationship between people and places (chapters 2-3). 5. Only now is the search on for phenomenological theories that can explain the clarified terms, phenomena, and concepts even better. I believe I have found such a theory in Hermann Schmitz's neophenomenological concept of atmosphere (chapter 4). 6. This concept of atmosphere is then examined and refined by reference to Judengasse conflict in Frankfurt – and enriched with Benjamin's concept of aura (chapter 5). 7. At this point (chapter 6), Schmitz's concept of atmosphere is deepened with his concept of situation – and it is shown to what extent this concept of situation enriches or could enrich contemporary discourse.

And when important methodological questions are raised in chapters seven and eight, they can be asked in a well-founded manner. They no longer hang in the air. This reduces, if not eliminates, the danger of a metaphysical-deductive approach. At the same time, it is understandable why a genius loci phenomenology cannot do without hermeneutics, and certainly not without field research and pathic-aesthetic correspondences (cf. chapter 7.3 „Christopher Tilley: 'make writing a voice for the stones, the places and the landscapes'").

What this phenomenological field research might look like and how it could culminate in a pathic dialogue is the subject of the *ninth chapter*, "Pathic dialogue with nature: Resonating, corresponding, poeticising. Pathic aesthetics in Bettina von Arnim and Ludwig Klages," following on from the fundamental distinctions made in the eighth chapter with Schmitz's concept of "aesthetic devotion." In this chapter, Tonino Griffero's project of "pathic aesthetics" is extrapolated in a romantic way, primarily through the terms and experiences of the incomparable Bettina von Arnim. Whether this extrapolation is productive remains to be seen.

1.3 *Author's note*

Chapters 2-7 of this book are based on texts that have already been published. Some of them have been presented at various conferences. I have thoroughly revised and updated all of these texts.

Chapters 1, 8 and 9 were written especially for this book.

Chapter 2 is based on my study "Der Geist eines Ortes. Eine kleine philosophische Kulturgeschichte des Genius Loci." In L. Mallien, J. Heimrath (2009) (eds.). *Genius Loci. Der Geist von Orten und Landschaften in Geomantie und Architektur*. Klein Jasedow: Drachen, 12-32.

Chapter 3 was published under the title "Genius-loci-bezogene Stadtentwicklung am Beispiel des antiken Rom." In H. Paetzold (2006) (ed.). *Integrale Stadtkultur*. Weimar: Verlag der Bauhaus-Universität 2006, 104-126.

Chapter 4 is based on chapter 7.3 of my 2004 monograph "*Der Geist eines Ortes: Kulturgeschichte und Phänomenologie des Genius Loci. Vol. 2: Neuzeit – Gegenwart.*" München: Albunea, 321-340.

Chapter 5 is the revised version of my article "Walter Benjamins Begriff der Aura und die Aura eines Ortes – am Beispiel der Judengasse Frankfurt." In H. Jung (2014) (ed.). *Symbolon, Jahrbuch für Symbolforschung*, Neue Folge, vol. 19. Frankfurt a. M.: Peter Lang, 193-217.

In chapters 6 and 7, I have partly drawn on the essay "Toward a Phenomenology of Nature-made and Human-made Genius Loci". *Environmental and Architectural Phenomenology*, vol. 34, no. 2, 2023, 17-24.

Without Tonino Griffero's invitation and motivation, this book would not have been possible: thank you very much for that, Tonino!

I would also like to thank Michael Großheim, the „Stiftung Neue Phänomenologie“ and the „Gesellschaft für Neue Phänomenologie“ for their financial and ideational support: thank you very much!

Last but not least, I would like to thank the translator Richard Falzmann: for his sensitive translation and for the philosophical discussions we had. To you too, dear Richard, a heartfelt thank you!

2.
Outline of a Cultural History and Phenomenology of the Genius Loci

Genius loci – the spirit of a place: a tradition-rich word for a long-standing phenomenon. From antiquity until today: this word emerges again and again, it is referred to on every occasion, we consistently want to tell something when using it, want to point to certain phenomena. This expression experienced a veritable renaissance, especially during the last 50 years in the course of non-receding debates concerning the environment and nature.[1]

---

1 Here is just a small selection of thematically relevant publications: J. D. Hunt, P. Willis (1975) (eds.). *The Genius of the Place. The English Landscape Garden 1620-1820*. London: Paul Elek. – Y.-F. Tuan (1976). *Geopiety. A Theme in Man's Attachment to Nature and to Place*. In D. Lowenthal, M. J. Bowden (eds.). *Geographies of the Mind. Essays in Historical Geosophy in Honour of John Kirtland Wright*. New York: Oxford University Press. – C. Norberg-Schulz (1980). *Genius Loci: Towards a Phenomenology of Architecture*, New York: Rizzoli. – J. Pieper (1984) (ed.). *Ort, Erinnerung, Architektur. Über den Genius Loci*. „Kunstforum" 69. – H. L. Garnham (1985). *Maintaining the Spirit of Place – A Process for the Preservation of Town Character*, Mesa: PDA Publishers. – M. C. Neddens (1986). Ökologisch orientierte Stadt- und Raumentwicklung: Genius loci, Leitbilder, Systemansatz, Planung. Wiesbaden-Berlin: Bauverlag. – U. Tworuschka (1994) (ed.). *Heilige Stätten*. Darmstadt: WBG 1994. – M. Pogačnik (1995). *Elementarwesen. Die Gefühlsebene der Erde*. München: Knaur. – C. Majunke (1999). *Der Genius loci: Geist des Ortes oder verorteter Geist. Landschaftsplanung zwischen dem Wunsch nach Ganzheit und moderner Subjektivität*. Berlin. – B. Pedroli (2000) (ed.). *Landscape – our Home / Lebensraum Landschaft. Essays on the Culture of the European Landscape as a Task / Essays über die Kultur der europäischen Landschaft als Aufgabe*, Stuttgart: Freies Geistesleben. – B. David, M. Wilson (2002) (eds.). *Inscribed Landscapes. Marking and Making Place*. Honolulu: University of Hawai'i Press. – R. J. Kozljanič (2004a-b). *Der Geist eines Ortes: Kulturgeschichte und Phänomenologie des Genius Loci*. 2 vols. München: Albunea. – G. Böhme (2006). *Architektur und Atmosphäre*. München: Fink. – L. Mallien, J. Heimrath (2009) (eds.). *Genius Loci. Der Geist von Orten und Landschaften in Geomantie und Architektur*. Klein Jasedow: Drachen. – G. Böhme (2017). *Atmospheric Architectures. The Aesthetics of Felt Spaces*. London-New York: Bloomsbury.

(Post)modern resurrection of the genius loci

In various disciplines – like architectural theory, human geography, art history, gardening and landscaping, but also in religious studies, ethnology, geomancy, archaeology, literary studies and last but not least in philosophy – at first sight the ever recurrent term of the genius loci creates the impression of having always to do with the same concepts, approaches and phenomena. But appearances are deceptive. It is a matter of different discourses, which always mean something else when speaking about the genius loci.

The spectrum of what is meant to be the genius loci reaches from the purely metaphoric and rhetoric meaning of the word over the historical meaning of a zeitgeist connected with a place and a socioculturally constructed "local spirit" to the probably original and closest meaning of the genius loci concept: the aesthetic and synaesthetic qualities of a place – above all the impressive "atmospheres" and shape-like "characters" of a place. There is talk of ecologically-systemic contexts as well; and last but not least of place-bound "energy fields" and local "nature spirits". What we have to (co-)consider almost always is: It has to do with a specific, evolved and inherent *quality*, which seems to be elusively related with the respective place – and which cannot be properly grasped and understood by common quantitative scientific methods. This is why a qualitative approach – a hermeneutic and neo-phenomenological one – is inevitable in my case.

Placelessness and topophilia

However, the various and partly contradictory meanings of the term genius loci seem, in my view, to underlie a common motive, a comprehensive need: to counteract the growing rootlessness of our time – a rootlessness, which Heidegger emphasised in his speech "Bauen Wohnen

---

– S. Pedone, M. Tedeschini (2017) (eds.). *Genius loci*. Milano: Mimesis (= Sensibilia 9/2017). – E. Arndt (2019). Der Genius Loci Lüneburgs. Erkundungen zu raumbezogener Identität. Lüneburg: IfSK. – T. Griffero (2020). *Places, Affordances, Atmospheres: A Pathic Aesthetics*. London-New York: Routledge, (Chapter "Found or created? The genius loci as an (atmospheric) experience", 137-149). – D. Seamon (2022). *Sense of Place*. In D. Richardson et al. (eds.). *The International Encyclopedia of Geography*. Hoboken: John Wiley & Sons. – J. D. Hunt (2022). *Genius Loci. An Essay on the Meanings of Place*, London: Reaktion. – R. J. Kozljanič (2023b). *Toward a Phenomenology of Nature-made and Human-made Genius Loci*. "Environmental and Architectural Phenomenology" 2, 17-24.

Denken" (Building Dwelling Thinking) in 1951 after all as placelessness (Heidegger 2000a, 145-164; Heidegger 1971, 143-161). Counteracting the growing homelessness and placelessness: This would mean the negative formulation. The love of one's native place or "topophilia" (amor loci, nexus loci, a sense of home and place) – to employ a term coming to light by Gaston Bachelard (1964) and Yi-Fu Tuan (1974): This would be a positive formulation for this motive and need.[2]

Starting from Heidegger the human geographer Edward Relph noted on this:

> If places are indeed a fundamental aspect of man's existence in the world, if they are sources of security and identity for individuals and for groups of people, then it is important that the means of experiencing, creating, and maintaining significant places are not lost. Moreover there are many signs that these very means are disappearing and that 'placelessness' – the weakening of distinct and diverse experiences and identities of places – is now a dominant force. Such a trend marks a major shift in the geographical basis of existence from a deep association with places to rootlessness, a shift that, once recognized and clarified, may be judged undesirable and possibly countered. It will then be of no small importance to know what are the distinctive and essential features of place and of our experiences of places, for without such knowledge it will not be possible to create and preserve the places that are the significant contexts of our lives. (Relph 1976, 6).

Well, whether we are talking about placelessness or topophilia, whether negation or position: as a matter of fact – these terms and feelings are filled with life and reality not before we correlate them with relevant living spaces or understand them anew from within. This is to happen now. This brief history of genius loci serves to lay the initial foundations. The task is to trace historical lines of development, to highlight specific spaces experienced in times past, and thereby to arrive at terms and concepts that are grounded in both cultural hermeneutics and phenomenology. This historical passage is a cursory and sketchy one in this book. I am more detailed and systematic in view of these contexts in my two-volume monograph: „Der Geist eines Ortes. Kulturgeschichte und Phänomenologie des Genius Loci" / „The Spirit of a Place. Cultural History and Phenomenology of the Genius Loci" (Kozljanič 2004a-b).

Now let us approach more closely the term and the phenomenon of the genius loci in order to determine altogether what was meant by this term when it was coined in antiquity and what kinds of experience were associated with

2 On the concept of "Heimat" („homeland") cf: Joisten (2012, 170-171).

it. Let us first cast a glance at the archaic and mythical origin of the genius loci. Then I will reproduce the undergoing transformations of the genius loci during the Middle Ages, the Renaissance, the Enlightenment and the Romantic era to finally end up in the phenomenological unconcealments of the 20th century.

2.1 *The mythico-archaical origin of the genius loci*

The term and cult of the genius loci in antiquity

The Latin term "genius loci" (pl. "genii locorum") literally means "spirit of place". According to the ancient Romans, every individual human being had his own genius. Similarly, *certain* places also had their own genius: in other words, a local spirit. It was believed that this genius constituted the *character* of a person or a place (Kozljanič 2004a, 28-78). This spirit was equally regarded as a protective or tutelary one, and thus sometimes called and honoured as the "tutela" or "tutela loci".[3]

"Spirit of place" thus means a place that has a special, characteristic spirit. A place has such a spirit either through itself, as a kind of innate spirit ("ingenium loci")[4] or through what people have inscribed in it, whether materially or immaterially. But whether innate or inscribed: decisive is that this spirit is an evolved one, one that has grown and become naturally and/or historically. Equally important: the genius loci is not only a spirit that has grown, but also one that continues to grow. For every genius – including that of a place – is a "nascent spirit", a "spirit of becoming" (in German: "Werdegeist"; cf. Birt 1886-1890, col. 1614; Kozljanič, 2004a, 50, 67). This is also suggested by the etymology of the word. "The origin of the name from the root gen in gignere is obvious and was not misunderstood by the ancients." (Otto , col. 1156). Gignere means to beget, to bring forth, to give birth to.[5]

---

3 See e. g. Petronius, 57, 2; or the inscription in the Corpus Inscriptionum Latinarum, full bibliographical details for which are available at: https://cil.bbaw.de/hauptnavigation/das-cil/baende (accessed June 24, 2025), and which is henceforth abbreviated as CIL; here, CIL, vol. 6.1, no. 216 p. 40; vol. 13.1, no. 440, p. 58).

4 Cf. Ovid, met. 3, 157-162, also: trist. 5, 10, 18; Pont. 2, 1, 52; 4, 7, 22.

5 According to the OED, the Latin *genius* derives from the base of *gignere*, 'to beget', and in turn the Greek *gignesthai*, 'to be born' or 'to come into being'.

These genius loci places can be roughly categorized into two groups: on the one hand into the artificial (humanly dominated) places being of social and existential importance, on the other hand the natural (dominated by nature) places having some sort of divine or better numinous aura.

With reference to the *artificial* place and its spirit we read at the Christian late-antique poet Prudentius:

> And yet why do I find you pagans imagining but one genius of Rome, seeing that you are wont to attribute to gates, houses, public baths, taverns, each its own genius, and throughout every part of the city at every point imagine thousands of geniuses, so that no corner shall be without its own ghost [shadow]? (Prudentius, *Symmachus*, II, 445-449)[6]

Of course, Prudentius exaggerates here a little – even for Christian and propagandistic reasons. As we will see later, not every corner has its genius, but only special places that are lifewordly relevant or historically important *characteristic* places.

With reference to the *natural place* and its spirit an interesting passage as Cato mentioned in his writings on agriculture. In this text, Cato explains the Roman rite that must be followed when thinning out a grove:

> A pig is to be sacrificed, and the following prayer uttered: "Whether thou be god or goddess [*Si deus, si dea es*] to whom this grove is dedicated [*quoium illud sacrum est*], as it is thy right to receive a sacrifice of a pig for the thinning of this sacred grove, and to this intent, whether I or one at my bidding do it, may it be rightly done. To this end, in offering this pig to thee I humbly beg that thou wilt be gracious and merciful to me, to my house and household, and to my children. Wilt thou deign to receive this pig which I offer thee to this end." (Cato, *De agricultura*, 139)[7]

A Roman wall-painting from Herculaneum (picture 1) shows in a very graphic way how the spirit of place was conceived and what it signified. In the painting we can see a small circular altar, around which a serpent winds its way upwards. Its tail touches the earth, its head stretches over the top of the altar. The serpent – unambiguously identified by the inscription to its right '*GENIVS HVIVS LOCI MONTIS*' as a manifestation of 'the genius

6 Thomson (1953). Latin text: „quamquam cur genium Romae mihi fingitis unum, cum portis, domibus, thermis, stabulis soleatis adsignare suos genios perque omnia membra urbis perque locos geniorum milia multa fingere, ne propria vacet angulus ullus ab umbra?"

7 Ash (1934, 121 = §139)

of this part of the mountain' – is eating the offering of food that has been placed on the altar.[8] To the left of the picture a boy is approaching the altar. In his right hand he is holding a branch. On his head he is wearing a garland. He could be a shepherd boy who, ritually dressed, is coming to this part of the mountain to make a sacrifice or offering to the genius for the welfare of his flock. This conjecture is supported by verses from Calpurnius. In spring the shepherd should first sacrifice – "Invoke with salted meal the gods, / The guardian genius of the spot / The Lars and Faunus" – then let the flock go to pasture (Calpurnius Siculus, 5, 26; Scott 1890, 95).

[Pic. 1: Roman wall-painting from Herculaneum, dedicated to "the genius of this part of the mountain".]

The corresponding term of 'genius' in ancient Greek is *daimon*. In the polytheistic antiquity daimon does not mean devil or satanic spirit – this meaning was only later attributed by Christian reinterpretation – but *daimon* simply means a numinous being, a divine spirit. There were many divine beings; some of them were place bound local daimons with the Greeks. Although the Greeks did not have a proper term for these local daimons – in contrast to the Romans – there was no doubt that special beings dwell special places. When, for example, in the Platonic "Phaedrus" dialogue Socrates says

8 See Birt (1886-1890, cols 1613-25; col. 1624 is a reproduction of the picture from Herculaneum).

in a nymph sanctuary situated in direct vicinity of the Illissos river – "for surely the place is holy; so that you must not wonder, I, as I proceed, I appear to be in a divine fury" (Plato, *Phaedrus,* 238 c-d)[9] – he does not mean anything else but the presumed presence of divine spirits and his exposure to their influence.

In summary: A genius loci was, according to the polytheistic worldview of antiquity, considered a numen or daimon, a daimonic deity. Usually, these local daimons were worshipped cultically. People prayed and spoke to them, sang and danced, consecrated something and sacrificed to them: that local life might be protected and promoted. Examples of such genius-loci-like beings are: mountain and river deities, spring and tree nymphs, hearth and store gods (lares and penates), forest gods and wilderness goddesses, local heroes and deified people, ancestor gods and spirits of the dead (manes), genies (genii) of the army and theatre, deities of the harbour or city. The classical anthropomorphic representation of the genii was standing with toga and cornucopia (cf. pic. 23). In the case of spring nymphs and river gods, they tended to be reclining with an overturned urn or vase flowing with water (cf. pic. 2). Theriomorphically, the local genii were often depicted as snakes, river gods sometimes as bulls. Tree nymphs were sometimes depicted in tree form (cf. pic. 20), mountain gods in mountain form.

[Pic. 2: Reclining nymph with an overturned urn.
Fountain fragment from Rome, 2nd century. Vienna Museum of Art History.]

9 Trans. by Benjamin Jowett. See: https://classics.mit.edu/Plato/phaedrus.html.

The mythico-archaical experiencing of the genius loci

Genius loci was not only a *term* being used in antiquity, it was not only a certain – more or less reflected, more or less widespread – view on local spirits. This term was above all connected to certain lifeworld and religious experiences and occurrences. Now I would like to turn to these aspects in order to eye more closely the mythico-archaical *phenomenon*, i. e. the lifeworld *appearance* of the genius loci.

The above mentioned Plato citation was already a hint: At certain places a numinous atmosphere is to be found, a certain secretively-divine sentiment, an impression, which can also cast a spell over you.

A description by Seneca (1st century A.D.) shows that the atmosphere of a genius loci site can be not only graceful, beautiful and lovely, but also sublime, eerie, oppressive, overwhelming and above all numinous. It is found in the 41st Letter to Lucilius:

> If ever you have come upon a grove that is full of ancient trees which have grown to an unusual height, shutting out a view of the sky by a veil of pleached and intertwining branches, then the loftiness of the forest, the mysteriousness of the place [secretum loci], and your marvel at the thick unbroken shade in the midst of the open landscape, will prove to you the presence of the numinous [numinis]. Or if a cave, made by the deep crumbling of the rocks, holds up a mountain on its arch, a place not built with hands but hollowed out into such spaciousness by forces of nature, your soul will be deeply moved by a certain intimation of the existence of the divine [religionis suspicione percutiet]. We worship the sources of mighty rivers; we erect altars at places where great streams burst suddenly from hidden sources; we adore springs of hot water as divine, and consecrate certain pools because of their dark waters or their immeasurable depth. (Seneca. *Ad Lucilium Epistulae Morales,* 4, 41, 3).[10]

I would like to name this experiencing as the *atmospheric experiencing of a place* – and will clarify it below (in chapter 4) with Hermann Schmitz' notion of atmosphere.

In order to give an insight into the deepest appearance layers of the genius loci – an insight into something that I will subsequently call the *daimonic experiencing of a place* – I want shortly turn towards to the Trophonios oracle in Lebadeia, Boeotia. Pausanias, in his "Description of Greece" (2nd half of the 2nd century A.D.) gave an impressive depiction of the ritual oracular inquiry of the cave daimon Trophonios there. According to Pausanias, the oracle seeker, after undergoing days of ritual preparations, (like fasting,

10 My translation (R.J.K.), based on: Gummere (1925, 273-275).

offering sacrifices, praying, ...) has to go to the cave of Trophonios at the favourable hour determined by exsticpicy and dressed in linen vest and local sandals. He lets himself slide into the narrow cave crevice feet first.

Let us listen to Pausanias at the most interesting passage himself:

> The descender lies with his back on the ground, […] thrusts his feet into the hole and himself follows, trying hard to get his knees into the hole. After his knees the rest of his body is at once swiftly drawn in, just as the largest and most rapid river will catch a man in its eddy and carry him under. After this those who have entered the adyton learn the future, not in one and the same way in all cases, but by sight sometimes and at other times by hearing. The return upwards is by the same mouth, the feet darting out first. […] After his ascent from Trophonius the inquirer is again taken in hand by the priests, who set him upon a chair called the throne of Memory which stands not far from the adyton, and they ask of him, when seated there, all he has seen or learned. After gaining this information they then entrust him to his relatives. These lift him, paralyzed with terror and unconscious both of himself and of his surroundings, and carry him to the building where he lodged before with Good Fortune and the Good Daimon. Afterwards, however, he will recover all his faculties, and the power to laugh will return to him. What I write is not hearsay; I have myself inquired of Trophonius and seen other inquirers. (Pausanias, *Description of Greece*, 9, 39)[11]

[Pic. 3: The cave of Trophonios near Lebadeia (now Livadia) was built over with a chapel ('Fountain of Life') in Christian times.]

11 Jones, Ormerod (1918). See: https://www.periegesis.org/en/reports.php?reportid=317

[Pic. 4: In the Hercynian Valley near Lebadeia/Livadia. These could be the springs mentioned by Pausanias from which the vision seekers had to drink.]

Genius loci and original polydaimonism – worldwide

All of these local deities, all of these genii locorum belong to the so-called lower mythology. This indicates not only the proximity of these religious ideas to popular belief, but also their great age. Belief in local protective deities is one of the oldest religious traditions in human culture. It is found worldwide in archaic (e.g. shamanic) cultures and can be traced back to prehistoric times.[12] It can also be found in neolithic-matricentric contexts.[13] Today's anthropologists and ethnographers also frequently come across these local spirits among indigenous peoples.[14] These local deities are based on numinous-atmospheric and visionary experiences of the place. As the natives credibly affirm, these deities can be bodily felt, mentally imagined and visionarily experienced as numinous or daimonic powers. One of the most profound interpreters of the primordial mythical worldview, the psychologist and philosopher Ludwig Klages (1872-1956), therefore speaks of visionary appearing "archetypes of reality"

12 Prominent here: the guardian goddesses of animals and caves; see Duerr (1985a, § 2f.). On Siberian shamanism cf. Findeisen, Gehrts (1983, 28-30).

13 Gimbutas (1991, 252, figure 7-45; "Guardians of Wild Nature").

14 Cf. David, Wilson (2002, chapt. 4); Mader (2002); or: Ingold (2020).

(“Bilder der Wirklichkeit” “primal images of reality”) and a closely related „polydaimonism“: “The true daimon is the daimon of a place, of an area, of an element, changing with their appearances, accepting the sacrifices of its devotees”.[15]

The mythico-archaical relationship of integration

If we want to describe what the relationship of the mythico-archaical human towards nature is, we could aptly speak of a relationship of integration.

A human integrates or subordinates into the naturally numinous place, which he encounters in the naturally grown, so to speak, long-established structure. The natural grotto will be a little gouged at best, namely to place in a cult image of the genius loci; the grove will be provided with an enclosure, namely to define it from the mundane exterior; a round little altar is set up on the mountain. But all creative interventions of humans subordinate to the principle of the discovered place, they dedicate themselves to the naturally numinous emanations and impressions.

2.2. *The Olympic-mythical structuring of the genius loci through cross-local deities*

The Olympic-mythical relationship of structuring (in terms of terminology and architecture)

The famous philologist, friend of Nietzsche, Erwin Rohde (1845-1898) writes:

> If we listened to Homer alone we should suppose that the innumerable local cults of Greece, with their gods closely bound to the soil, hardly existed. Homer ignores them almost entirely. His gods are pan-Hellenic, Olympian. [...] In his picture Greek beliefs about the gods *appear* absolutely uniform [...]. In reality—of this we may be sure—no such uniformity existed; the main outlines of pan-Hellenism were doubtless there, but only the genius of the poet can have combined and fused them into a purely imaginary whole. (Rohde 1925, 25f.)

15 Cit. in: Bishop (2018, 101f.); Klages (1969, 1264).

The Olympic world view with its Panhellenic Gods and Goddesses has now formed a peculiar alliance with the traditional archaic popular religiosity and the tradition of local cults. In informal analogy to Nicolai Hartmann's theory of strata – we could speak of a *relation of structuring* ("Überformung"; cf. Hartmann 1954, 122).[16] Here structuring expresses the transformation from the older stratum to the younger one without losing anything substantially. Thus because – still referring to Hartmann's theory – there is the same categorical structure in both strata – the mythological one in our case. Olympic world view and archaic popular belief are equally mythically structured: this is why they can correlate with each other at any time. To put it even more clearly and in my own words: by 'structuring' I mean that the previous, evolved basic structure is expanded and restructured – but without completely deconstructing it. The restructuring leaves the decisive elements of the basic structure intact, takes them with it and embeds them in the overarching structure. This means that the existing landscape is reshaped but not deformed.

The numerous antique temples consecrated to Olympic deities are an obvious example.

In the interest of brevity, let us refer here to the Hera sanctuary of Samos where this relational transformation is clearly evident. Instead of a daimon (poss. a tree nymph) attached to a place (cult tree) we see the scenically overarching goddess (Hera) and her temple. And the like applies for the Zeus sanctuary, near Dodona with its holy oak tree. Or for the Apollo sanctuary near Delphi with its adoption and transformation of several archaic genius loci sites: the Gaia rock, the omphalolith (Omphalos) the two springs of Kassotis and Castalia.

16 See also chapter 8: 'The Strata Laws of the Real World' in Hartmann (1953, 73-83). Unfortunately, the terms 'Überformung' and 'Überbauung' are translated somewhat one-sidedly as "superinformation" and 'superimposition'.

[Pic. 5: The gorge with the spring of the nymph Castalia, below the sanctuary of Apollo at Delphi]

The Olympian-mythical experience of landscape gods and landscape numena

It was consistently noticed that a good deal of antique temples have a clear environmental connection; the on hand they adapted to the surrounding scenery, on the other hand they emphasised the characteristic features of this environment, we may say they got to the heart of it. Here are a few examples: A temple of the so-called cow-eyed Hera is often connected with to humid, cattle rich lowlands (Argos, Samos, Poseidonia, …) suitable for cattle husbandry. The more remote temples of the so-called "far-shooting" and sunlike Apollo prefer light hillsides with a distant view (Delphi, Ptoon, Bassai,...). The large temples built in the valley to the thundering and lightning-throwing mountain and storm god Zeus (e.g. in Dodona and Olympia) are mostly connected with pretty high cloud-collecting mountains with smaller altars erected on top of them (e. g. Tomaros and Lykaios, also Olympos, Ida, ...). Sometimes there is even a temple directly on the summit (Garizim).

[Pic. 6: The classic landscape of Hera of Argos with its cattle-rich lowlands. The view extends beyond the Temple of Hera to the Argivian plain.]

[Pic. 7: Apollonian landscape in Delphi. View over the Temple of Apollo. On the left are the Phaidriades, below is the Castalian Gorge.]

The classical scholar Paula Philippson pursued these references in her work "Griechische Gottheiten in ihren Landschaften" ("Greek Deities in their Landscapes"). She could confirm and extend them. The landscape surrounding the temple was described as the landscape of cult of the corresponding God. If the temple is the culminating point of this landscape, hence the landscape is the essential appearance of the respective God (Philippson 1939). Vincent Scully went further along this path. In his work "The Earth, the Temple and the Gods" (Scully 1962), he pointed out that temples are not simply dwellings of the god or goddess, but have the task of constellating the landscape in such a way that the divine forces within it can be more easily and clearly grasped.[17] One could therefore speak of *landscape numina*, landscape-bound deities or, in short, *landscape deities*.

I have already referred to the Heraic, Apollonic and Zeus-like temples and cult landscapes. I would also like to give some particularly characteristic sanctuaries for other Olympic deities. This will give you a brief idea of the respective areas and the characteristic cult landscapes of these deities.[18]

- *Poseidon – the sea god* (lat. Neptune): The famous *Poseidon Temple at Cape Sunion* conveys a tremendous impression of what a Poseidonic landscape can mean. Equally quite obvious are the Poseidon temples of Isthmia and at Tainaron. 'Poseidonic' by all means is the temple and the landscape of the goddess Aphaia of

17 The starting point in Vincent Scully's work is the Minoan-Mycenaean mother goddess. His investigations are based on this important matriarchal, but also pre-Olympic concept. And it is against this background that Scully basically interprets all the Olympic temples and cult landscapes he deals with. This method remains fruitful to a certain extent, but only to a certain extent. When reading his work, I at least kept asking myself whether the unique nature of the Olympic deities (including their architectural and landscape characteristics) is not obscured, misjudged and generalised if, like Scully, they tend to be understood as deficient and faded late forms of a single basic pattern - namely the great mother goddess and her male hero. In my opinion, this measures the Olympic-mythical religion with incompetent categories. Paula Philippson takes a different approach. She deliberately remains within the realm of Olympic-mythical ideas and phenomena, without completely ignoring Minoan-Mycenaean influences. My investigations are therefore primarily orientated towards her approach and the results of her work. I also read Scully's work from her methodological understanding and perspective.

18 I have analysed and presented the landscape deities in more detail in: Kozljanič (2004a, 149-269).

Aegina, who is closely related to Poseidon (cf. Schoder 1975, 16f., 97-99, 198-201, 226-229; Scully 1962, fig. 283f.).

- *Hades – the god of the underworld* (lat. Pluto): Hades sanctuaries bearing witness of a Hades like landscape are the temples of *Zeus Katachtonios at Lebadeia*, the temple of Hades (the so-called Plutonium) in the sanctuary of Demeter at Eleusis with the subjacent Hades grotto (entrance to the underworld), the Plutonium and the incubation sanctuary at Acharaca near Nysa on the Meander river with an underground cave. The sanctuaries of Demeter and Persephone in Thera are also to be found in a higly Hades-like landscape (cf. Philippson 1939, fig. 12b, 14b; Schoder 1975, 62-64.; Scully 1962, fig. 114, 124).
- *Demeter – the Greek goddess of grain* (lat. Ceres): The paradigmatic Demeter landscape is characterized, of course, by the *main sanctuary of Demeter*, where her principal myth is present: Eleusis, the cropland of Attica (cf. Schoder 1975, 62-64). However the temple of Demeter at Akragas and the Demeter sanctuary in Ancient Corinth have to be taken into consideration as well (cf. Scully 1962, fig. 121f.; Schoder 1975, 121f.).
- *Artemis – the goddess of wilderness and hunting* (lat. Diana): The *Artemis temples of Brauron and Ephesus* as well as the Artemis-Cybele sanctuary in Sardis unveil an impression of what an Artemis landscape may look like (cf. Scully 1962, fig. 149f; Schoder 1975, 234-236). In her wildest wise Artemis reveals herself in and around her sanctuary in the mountain named after her Artemision and lying between Argolis and Arcadia.
- *Aphrodite – the goddess of love* (lat. Venus): An Aphrodite landscape represents on a large scale, with many exponentiated Aphrodite landscape parts on a diminished scale, the *island of Kypros* (Cyprus), which was *holy to Aphrodite* and was considered as extremely fertile in antiquity. The Roman Venus temple in the bay of Baiae at the Lacus Lucrinus (lake in Campania) also offers a significant scene of a lush Aphrodite landscape.
- *Athena – the goddess of defence, handcraft and the town* (lat. Minerva): *Athena* and her cultic landscape reveals herself most formidably at the *Akropolis in Athens*; and not only there, but in countless other Akropolis-like sites, where an Athena temple is enthroned above a city turning the surrounding landscape into a centred and clearly structured one (like in Troizen and particularly impressive in Lindos) (Scully 1962, fig. 320f.)

- *Dionysus – the god of wine and ecstasy* (lat. Bacchus): Dionysus may have revealed himself scenically in the colony of Miletus called *Dionysopolis* famous for its viticulture and also in Thracian Ismaros renowned for its wine in antiquity. (cf. e.g. Homer, Od. 9, 40 and 45; Virgil, Georg. 2, 37).[19]
- *Helios the sun god* (lat. Sol): An impression of power and presence of Helios in the landscape is conveyed by particularly holy places like *Rhodes* and also by Syrian Heliopolis (Baalbek) and Emesa (cf. Schoder 1975, 181-184).
- *Tyche – the goddess of fortune and fate* (lat. Fortuna): She reveals herself and her landscape in Praeneste.[20]
- *Asklepius – the healing and medical god* (lat. Aesculapius): He reveals himself in his mild, pleasantly relaxing, salutary landscape particularly so in *Epidaurus* and *Kos*.[21] Also the Amphiareion is to be found in an Asclepiadean scenery, which shows many similarities with the Asklepius sanctuaries (cf. Scully 1962, fig. 397f.; Schoder 1975, 22-25, 64-69, 124-127).
- *Pan – the wilderness god and god of shepherds* (lat. Faunus): He reveals himself in the rough mountain scenery of *Arcadia*. He has an important sanctuary (holy grove) at the mount Lykaion, beneath the Zeus altar (cf. Pausanias 8, 38, 5) and – spread all over Greece – countless cult caves, which are often located in wild and lonely mountain landscapes.

19 Cf. the characterisation of a Dionysian landscape by Ausonius in: Kozljanič 2004a, 242f.

20 Cf. the detailed description of the Fortuna sanctuary at Praeneste in: Kozljanič 2004a, 260-263.

21 Cf. the description of the Epidaurian atmosphere in: Nebel (1983, 84): „I enter the sacred area [...], I feel mysteriously moved, no tension as in Delphi, but a cosy relaxation, and I ask myself what is experience here, what is illusion. As if in waves, drowsiness spreads from a centre beyond myself. One more hardness must be softened, one more resistance dissolved, I let go of the future and glide trustingly into the Epidaurian now, which takes me in as if with cushions and pillows.“

[Pic. 8. Sunset at Cape Sounion. Poseidon Temple overlooking the sea. Here, the 'Poseidonian' can be experienced.]

[Pic. 9. Wild gorge in the Attic hinterland, north of Fyli. Almost at the bottom of the gorge lies a lonely Pan cave.]

All these examples show that the Olympic and mythical Greek had a clear consciousness of scenic realities; he was not only able to experience and to seize the atmosphere of a landscape but he engaged himself into it with the architecture of his temples, he illustrated, emphasized, enhanced it. I am therefore speaking here of *an atmospheric experience of a numinous landscape* (taking into account the atmospheric and daimonic experiences of places occurring in it). It is obvious that this experience is *similar* to the aesthetic experience of landscape in modern times; but that it is not *identical* with the profane, often purely superficial aestheticism of a pleasurable experience of modern times should be more than clear from the mythical context in which the Olympic-mythical experience of landscape is situated.

2.3 *The Christian-medieval schism of the genius loci into a local demon and a local saint*

Theological, moral and conceptual transformations

With the end of antiquity, i.e. with the rise of Christianity, the assessment of the mythical-mythological genius loci concept changed. It is above all the Judaeo-Christian monotheistic *idea of creation*, which makes the genii locorum appear as derived and seceded phenomena. It is Christian eschatology that also makes the genii locorum appear to be in need of redemption and ultimately dissolution. And Christianity's total moralisation of all dimensions of reality which captures the traditional concepts of deities and spirits are cut into two, by means of the logically-moral phrase of the barred third way – either good or bad, a third way is not permitted: Here is the omniscient gracious God with all his good spirits (the angels), there the overall evil devil with his vicious spirits (the demons).

In the process it is important to retain: The daimonic reality of local spirits is not questioned despite all these tendencies, only its assessment changes. The ambivalent and divisive relationship of medieval Christianity towards the genius loci does not only manifest itself in the history of pilgrimage sites – the holy places of Christianity – but also in folk religion and superstition with its uncountable Satanic places. This is why a more or less accepted local cult transformed into Christianity establishes itself at pilgrimage sites. The Satanic places, however, is a demonised cult of genius loci, which is inacceptable from a Christian point of view. Demonised places become Satanic places: Devil's Mountain, Devil's Rock, Devil's

Tower, Devil's Ravine, Devil's Table, etc. Divine daimons do no longer dwell there but *local demons* – "little devils", so to speak. "Baptised" places become pilgrimage sites, i.e. places of miracles and grace. For example, an ancient spring sanctuary becomes a place of pilgrimage with the name "Mary's spring of life", "Mary's well", etc. The old genius loci sites were no longer used to worship daimons or local gods, but local saints: saints with a more or less strong local connection; saints who were then canonised in medieval collections of legends, e.g. the "Legenda aurea" by Jacobus de Voragine (1228/29-1298).[22]

[Pic. 10: The blessed Edigna of Puch. In the background, a pagan place of worship with a lime tree, later built over with a chapel. Copperplate engraving from the collection of legends "Bavaria Sancta" 1615/27.]

The Christian superstructure relationship

Let us have a look at the Heraion of Samos at the end of antiquity. The mythical age draws to its end, the Christian one dominates. Christian "logos" overruled the myth ("mythos") and uses it as material to build

22 For more details, see Kozljanič (2004a, 271-401).

its new realm. Christian zealots like Firmicus (4th century) follows the maxim: "Beat them up, beat them with an axe, these temple ornaments! To the smelter, to the mint with these deities! All votive offerings are yours, take and use them" (cited in Burckhardt 1954, 300). The sanctuary of Hera on Samos with its tradition of three thousand years is destroyed by such zealots. The ancient cult tree is chopped down by Christian hand. The sacrificial fire no longer burns on the altar; not one stone is left upon the other. Directly beside the ancient place of cult an Early Christian basilica is built from the rubbles in the 5th century. Like most churches, it is aligned to the East, where the sun arises, the symbol of the Christ-logos. The concrete connection to the place is loosened, the cult tree is not worshipped for a long time, the genius loci is not very relevant (cf. Walter 1990, 199f.).

We cannot speak of a relation of structuring here. We should rather speak – again in a loose analogy of Nicolai Hartmann's theory of strata – of a relation of superstructuring ("Überbauung" – cf. Hartmann 1954, 124f.). For one thing is clear – and this is the difference to the relation of structuring ("Überformung") – that in the transition from the mythical layer to the Christian one, essentials are lost, new things are added. The Christian process of superstructure is based – both architecturally and ideologically – on a prior deconstruction of the mythical structure. The debris of what was previously deconstructed is then reused as set pieces that have been torn from their original context and alienated.

It is no longer about the same categorial structure: The Christian logos is categorially differently structured than the antique myth: it is noetic and transcendental, i.e. mentally structured – no longer spiritual-sensual. It is not the "spell of the sensuous" (Abram 1996) that should be heard and interpreted, but the mental-extrasensory call of the supernatural – the transcendence. The fundamental categories of the Christian spirit, however – just to mention a few – are: Overcoming the world, radical transcendence, enmity of the senses and body (accompanied by enmity of images and place), and polarising monotheism. They confront the mythical categories diametrically. It would seem to speak about a relation of superstructuring because the Christian relationship to the world is strongly dominated and determined by a theological, spiritual, and moral superstructure. And to put it concretely, the fact that many churches were built on pagan places of worship – overlaying and remodelling them – also suggests a relationship of superstructure.

There is no relation of superstructuring that totally breaches with all aspects of older layers. Hartmann also points to this fact. His words can be transferred here directly into our context. "Thus a whole series of categories breaks through relation of superstructuring turning it into a relation of structuring on a small scale, but on a small scale only, for the relation of transformation demands that all categories pervade. And this is not the case here." (Hartmann 1954, 125).

In our example it means that the basilica was built at the same place after all, thus preserving the continuity of place; it was even built with the same stones as its predecessor temple. And possibly – which is the case of many churches – a saint/a saintess takes over the role of a local deity; maybe the church becomes a famous pilgrimage church and the religious parade takes over the role of earlier cultic processions.

Christian follow-up cult: From local daimons and landscape deities to local saints

In many cases a follow-up cult resp. succession can be reconstructed. For instance Elijah, the biblical prophet pertaining to the Old Testament, became the classical successor of many ancient Zeus, Apollo and Helios places of cult, which were mostly situated on hills or summits. Such a procedure opened up as there is a tonal similarity Helios Elijah on the one hand; and on the other hand there is a connection of the Elijah legend to a mountain (Carmel) and a sign of God (a thunderbolt sent by Jahveh).

How complex the question of a follow-up cult and the succession of place of local deities is, can be demonstrated by the practice of a "high places" cult also in the Christianised West. "Generally the Latin West was reserved towards an official worship of Old Testament saints," according to Theofried Baumeister. Therefore Elijah was never distinctly worshipped in the Roman Catholic world. And yet a similar cult of "high places" can be observed here. It is tightly connected with the name of St. Michael the Archangel (Baumeister 1991, col. 1008). This is why we can speak of a veritable *St. Michael cult of high places.* An early major St. Michael sanctuary can be found in a grotto on the Apulian mountain of Gargano, also called Holy Angel's mountain, Monte Sant' Angelo. Its foundation dates back to the 5th century and was caused by St. Michael's three epiphanies. The last epiphany happened on September 29th, 493. September 29th is still considered St. Michael's day today. The grotto was earlier sanctified to Apollo. Until today the pilgrimage site enjoys great popularity (Baumeister 1991, col. 1011f.; Kriss, Kriss-Heinrich 1955, 192f.)

[Pic. 11: Michaelsberg near Gundelsheim with St. Michael's Church at its summit. St. Michael and St. Mary succeeded Jupiter and Juno Regina as cult figures here.]

[Pic. 12: Michaelsberg near Gundelsheim: Altar of Jupiter and Juno Regina. Now attached to the outside of the church.]

East of Capua, on Tifata, another Italic mountain, there was a Diana Tifatina temple and a Jupiter Tifatinus temple. From the former rubble arose a Benedictine monastery with the church Sant' Angelo in Formis (Radke 1979, cols. 823f.; Preller 1997, 146). Here a Jupiter and Diana mountain was christened to an angel mountain.

A quite important figure of saint, actually the most important one, is the Mother of God, the Blessed Virgin Mary. No figure of saint has absorbed so many different deities. The veneration of Mary is something special, not least in this respect.

> Religious studies saw and like to see in Mary a successor of Astarte, Kybele, Artemis of Ephesus, Hera (Juno), Isis, Ceres (Demeter), Venus and other mother goddesses or holy-unholy virgins. Many quite popular medieval legends of Mary, among them rather dubious ones and contradictory to modern sentiment indeed, according to catholic scholars, certainly have facilitated the replacement of pagan figures by Mary." (Wrede 2000, col. 1649).

In many cases Mary adopted the place of the city goddess Athena resp. Minerva, resp. Roma, such is the case of the Parthenon in Athens, as well in Rome, where the church Santa Maria Nova requisitioned the Western parts of the double temple Roma-Venus. The church of Santa Maria sopra Minerva, situated in straight proximity of the Pantheon, tells by her name that it was erected on a Minerva sanctuary. Yet Mary was not only the cultic successor of Athena/Minerva, but also of a quite different goddess of landscape: Artemis of Ephesus. The Church of St. Mary of Ephesus is one of the very early Churches of St. Mary. According to legend, Mary was said to have had her last abode on earth there (Wrede 2000, col. 1640). Artemis, Athena and Minerva as well as Mary are worshipped as holy virgins: So this seems all too natural to let the old mythical beliefs drift on the level of virginal similarities to the new time of Christianity. However, it is obvious that such an analogy remains on the surface. Mary is a virtuous virgin and Mother of God: Artemis, Athena, and Minerva are bellicose virgins and besides no mothers: an essential difference for which the Christian process of transformation has little regard. Its goal is a different one. It is only interested in the polydaimonic and polytheistic myth in order to overcome it.

Mary is the best example for the Christian way to occupy old mythical sites by Christian local saints. The Christians were fully satisfied when the new holy figure had an outer similarity in few facets with the previous mythical figure. At least in fading late antiquity too many similarities – when mythical cult was still widespread – were rather annoying: One had to justify oneself towards “pagans”, had to explain that the adoration of saints did not mean at all worshipping gods and idolatry, but something completely different. This was apparently the more difficult the more similarities occurred. Concerning Mary it is obvious that she is not so locally and landscape bound in the end. She can basically take the place of any mythical goddess. Through the transmundane connection her concept is basically extensible to all places and landscapes like the concept of all the other saints as well. It bursts the antique complex of ideas concerning local daimons as well as that of antique landscape deities.

The transition of the mythical notion both of the local daimon and/or the landscape deity to the notion of the local saint can be decompounded in summary by the following steps.

First: The presence of a mythical genius loci figure or that of a landscape deity with the corresponding cult.

Second: (False) attribution of a historical or pseudo-historical Christian holy figure.

Third: Endowment of this saintly figure with a cult legend associated with her, which installs her Christian, mostly ‘monastic’ virtues (confessional courage, faithfulness to Christ and endurance of torture and suffering; humility, poverty and virginity; hatred of daimons, topophobia and fixation on the afterlife) as absolutely exemplary and worthy of imitation.

Fourth: Adoption of some ‘accidental aspects’ of the original mythical figure.

Fifth: Reinterpretation of the ancient mythical place cult and the adoption of some of its aspects transformed into Christianity.

Sixth: Deportation of the original genius loci figure into the realm of demons and/or into the realm of oblivion (and repression).[23]

The Christian-medieval experiencing of the genius loci

In its last consequence the Christian experiencing of a place was vertical, hierarchically transcending upwards, orientated to the Lord and Creator. This experience of place, in Christian-Platonic terms, tended towards the supra-heavenly place; in contrast to the mythical experience, which was more horizontal, immanent and worldly.

But only: *as a last resort*. Before the last resort the medieval experiencing of the place was comparatively mythical, as it reflected in the sagas of local demons and in legends of local saints.

23 I take essential elements of the above list from: Kufner (1992, 159).

[Pic. 13: The legendary water nymph Melusine.
Woodcut from Thüring from Ringoltingen's "Melusine" from the "Book of Love" (1587).]

I say: *comparatively*, for it was superstructured in a Christian way, which can be seen, for instance, that also the sagas and even more so the legends are mostly Christian morally and dichotomously oriented: Local demons and landscape deities turned either into evil doom-yielding devils or good local saints. Accordingly, the daimonic experiencing of the place split into a demonic experience of the place in legendary, tendentially 'diabolical' places there and a miraculous, wonder-working experience of the site in 'holy' places of pilgrimage and grace there.

A stark example, which is quite enlightening not least because of its similarity with the Trophonius oracle, originates from the Augustinian monastery and the local St. Patrick's cave in the Irish County Donegal. A bipartite island lies in a lake there. One part is pretty prolific and graceful the other one rather wild and bleak. There is an Augustinian monastery in its first part, in the second part a cave also called "St. Patrick's Purgatory".

According to an Irish legend St. Patrick was led by God into this lonely region and shown this cave with the following advice:

> those who enter it well prepared through [days of fasting and prayer and] sacraments abiding there day and night will obtain his purgation; and his sins will be forgiven, whereas the unrepentant will perish in it. The saga adds: Some who visited it never returned; but those who had returned remained faithful to the faith; but had laughed never more; because what they saw there embittered them the lust of the world.

A certain Oenus is

> taken to the cave around the year 1152 [after the ritual preparations] in the usual way, confined there and has to perambulate ten consecutive places of torment. What pests and tortures imagination can concoct is to be found here; some are entwined by firedrakes, others suspended above sulphuric flames and others plunged into baths of liquid metal; while some cling pale to a rock summit as if awaiting death or worse until a tempest seizing the deplorable ones hurls them into an ice-cold, stinking stream flowing past, in which the perambulating knight would almost perish had he not been saved by calling on the saviour. (Görres 1989, 92f.; cf. Duerr 1985a, 327f.).

Another, more beautiful example that shows the profound and loving way in which it was (and is) possible to interact with and experience the genius loci under Christian auspices is the famous Catholic pilgrimage site of Lourdes. Here the Blessed Virgin Mary appeared in a grotto in 1858: quasi a Christian glorified cave goddess and spring nymph. Bernadette Soubirous, a 14-year-old peasant girl, had several Marian apparitions and was later canonised by the Catholic Church. Admittedly we are leaving hereby temporally the Christian Middle Ages, but not topically. For long periods Lourdes is a late flowering of miraculous and medieval experiencing of place.

[Pic. 14: Lourdes in 1858. Saint Bernadette sees the Mother of God appear in a grotto. A healing spring (below) rises at the site of the apparition.]

[Pic. 15: Lourdes is one of the most important pilgrimage sites in Catholic Christianity. It is clear to see that the grotto is dominated by the church built above it.]

2.4 *Modern era transformations of the genius loci*

Aestheticisation, reformation and rationalistic reductions of the genius loci

With the beginning of the subject-centred and rationalistic modern era further decisive transformations are associated with the traditional concept of the genius loci: It comes to aestheticisation, to reformation and also to rationalistic reductions and dissolutions.[24]

When asking how the modern era transformations of traditional genius loci concepts took place in particular, we can discern the following steps:

First, in the Age of Humanism and the Renaissance (from mid-14t[th] to the end of the 16th century): Here comes the transforming rebirth of the ancient genius loci concept. This rebirth is transforming because it regards the ancient genius loci concept over long periods from behind a Christian screen and – what is more important – because it rediscovers it under modern era subjectivist and modern era aesthetic conditions. The

24 The following is a very brief overview and summary. I have explained these points in more detail in: Kozljanič (2004b, 9-130).

place receives its 'spirit'" through the artistic subject, which – through its individual reaction to the place – transfers to it certain spiritual and aesthetic significance, which are not absolutely due to the place as such, but first of all due to the subject. Thus the genius loci becomes an aesthetic metaphor of the subject. In the process of freeing itself in these times gradually from the sacred realm/sphere, a central role is devolved on art provided it "metaphorizes" nature through "aestheticisation": Daimonic and demonic places and landscapes become *metaphorical places and landscapes*. The result is a specifically modern, atmospheric, aesthetic and pleasurable experience of picturesque places and landscapes. – I cannot accomplish this here *in extenso* but I still want give a few keywords: Picturesque gaze that also 'frames' the landscape; aesthetically-distanced, enjoying look on the horizon; central perspective and vanishing point; gradation of the background by means of blue and haze shading. – Important humanists of the Renaissance like Francesco Petrarca (1304-1374) and Aeneas Silvius Piccolomini resp. Pope Pius II (1405-1464) paved the way for this new profane, aesthetic and picturesque experiencing of the landscape and then, of course, the landscape artists of the 15th to the 17th century.

Second, in the Age of Reformation: There is a first wave (since the 16th century) with Luther (1520 three church reform epistles), Zwingli and Calvin and their rejection of the Christian medieval doctrine of the Saints and the cult of pilgrimage. The 'purification' of churches from saints and cultic pictures, holy graves, relics is followed by the profanation of holy sites. In a second wave (since the end of the 17th century), the medieval demonology is principally opposed by the Amsterdam based preacher Balthasar Bekker (1634-1698) and his work "De betoverde wereld" (1680; in English: "The bewitched world"). In both cases the autonomy and reality of local saints and local daimons is denied from the viewpoint of the rational Christian subject. Both types of local sprits express inadmissible and debunkable externalisations of the subject. With the local saints and their pilgrimage cult, the only appropriate and tolerable way of communicating with God – in the eyes of the reformers – was externalised and distorted: namely direct personal and unmediated inner contact with Him. In the case of local demons, diverse delusions of the subject are externalized. In both cases the place receives its 'spirit' through a subjective deed of outward relocation resp. projection.

Third, the philosophers of the Enlightenment of the late 17th and 18th century: Here the tendency prevails to deny the genius loci (no matter whether in the antique form of local daimons or in the Christian form as local demons or spirits) as well as other mythical and religious figures in their character as independent, distinct phenomena. They deem to debunk them as derived phenomena, as a form of self-deceit and/or fraud committed by others. E.g. the French „*Traité sur les trois imposteurs*" (in English: „*Treatise of the Three Impostors*"), which are Moses, Jesus, Mohammed; written by an anonymous author towards the end of the 17th century; and the writings by Bernard Le Bovier de Fontenelle (1657-1757) "*De l'origine des fables*" (1684 – in English: "*On the origin of myths*") and "*Histoire des oracles*" (1687 – in English: "*History of Oracles*").

The exposure of the Enlightenment makes use of four, partly explicit partly implicit, various reductionist argumentative strategies: the evolutionary, historical, social and psychological ones. The first argumentative strategy tries to interpret the mythical concept of the genius loci in an evolutionary way as an early 'childhood' stage of human evolution. Hereby the myth-believing early man is imagined being like a child that believes in all sorts of unreal fairy tales and epitomes of its own imagination. The second argumentative strategy tries to undress the myth of its envisioned imaginations in order to progress to the authentic 'true' *historical* core of it; for many myths, they think, contain an element of historical truth. The third argumentative strategy tries to reduce the genius loci to its *social* core, to its social function. The fourth argumentative strategy, which is most influential over longer periods, explains the genius loci psychologically as a projection of early man's psyche into nature. Thus early man, living without the enlightened, scientific explanations could only explain all those frightening, astounding, powerful, natural phenomena by spiritual fantasies and phantasms of gods.

All four enlightened argumentative strategies can be found intertwined under the omen of the "priestly fraud thesis" in the Great Encyclopaedia of Diderot and d'Alembert (18th cent.). As Diderot put it: "Ridiculous superstitions, silly, sometimes abominable mysteries, fantasies and lies designed to consolidate the authority of the priests and intimidate the blind crowd."

The experiencing of the genius loci in the new era of rationalism resp. geometry

For a strictly rationalistic human like *Descartes* (1596-1650) the world is split into two substances: a mental substance (res cogitans) and a corporeal substance (res extensa). Everything that does not think, which is not a thinking human (God as the supreme mind certainly excepted) is corporeal substance. Also plants and animals (and even man, but only inasmuch as he also is a corporeal being and not a thinking being) are machines made of soil. The world is a machina mundi (world machine). In the beginning of the world the supreme mechanic, namely God, assembled and wound up the whole mechanism in order to start it then. Then the course of the world is conceived like a wound up clock with its purring wheels.

When such a Cartesian thinking person looks into the world, into nature and the landscape, he sees only inanimate matter. As it is inanimate and therefore quasi senseless and shallow, it is needless to ask for a spirit of the place. It cannot be experienced out there unless a human has imposed his spirit, i.e. the rational order of his spirit on nature. This implies that places only acquire spirit, soul, character through humans. Behind all this is a new understanding of space: the geometric, three dimensional; an understanding of space that is accompanied by a peculiar, new experiencing of space: a mathematically constructing and reconstructing experience of space.

The modern era (re)discovery and exploration of geometric space took place before the 'Cartesian reduction' – it took place even before the 'Copernican revolution' (1543). Renaissance artists such as Filippo Brunelleschi (1377-1446) and Albrecht Dürer (1471-1528) played an important role in this. They attempted to see space in an abstract, three-dimensional way in order to be able to reconstruct it in two dimensions by means of an experimental set-up and with the help of drawing apparatuses. I will only briefly mention Brunelleschi's famous experiment in 1425 at the Baptistery of San Giovanni in Florence, through which, it is said, he discovered linear central perspective (more on this in: Feyerabend 1984, 17-23). In 1525, Albrecht Dürer published a book entitled "Instruction in measuring with the compass and straightedge in lines, planes and whole objects compiled by Albrecht Dürer and printed for the benefit of all art lovers with corresponding figures". It is the first summary of the mathematical-geometric methods of central perspective. This laid the foundation for the perspective-geometric experience and (re)construction of space in art and science.

The modern geometrical relation of excessive planning (overplanning)

One example that clearly illustrates the Enlightenment approach to nature and shows that this period (in its most reductionist representatives) was not really interested in a pathic experience of nature is the French garden; namely exactly there where it was realised in its purest from, at Versailles.

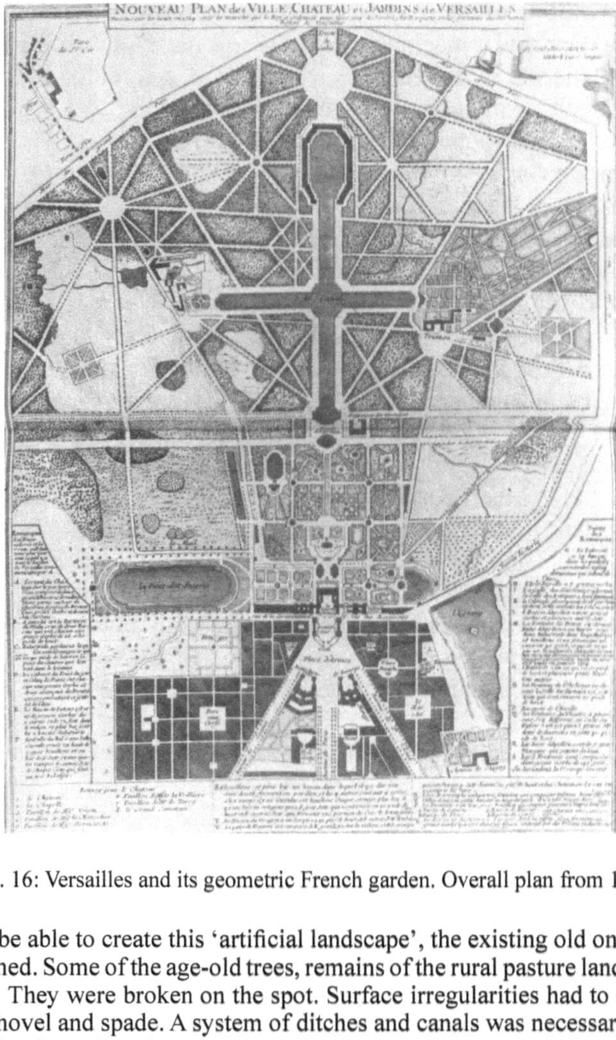

[Pic. 16: Versailles and its geometric French garden. Overall plan from 1714]

To be able to create this 'artificial landscape', the existing old one had to be abolished. Some of the age-old trees, remains of the rural pasture landscape, had to fall. They were broken on the spot. Surface irregularities had to be levelled with shovel and spade. A system of ditches and canals was necessary to ensure the irrigation of the huge gardens. For forty years an army of gardeners and workers were shaping these grounds caring and rebuilding them. [...] Saint-Simon, a critic of Louis XIV, calls it 'the arrogant pleasure to impose nature his will'. [...] Not only the free growth of plants, the unimpeded flow of waters and the naturalness of the landform were subjected to man, to the king, also the animals had to bow to his will. Le Roi Soleil presented it with his 'Ménagerie royale de Versailles'. In animal enclosures, designed under aesthetic and architectural aspects and not under animal needs, the 'beasts' were here on display. (Buderath, Makowski 1986, 23f.)

[Pic. 17: Schönbrunn Palace, Vienna, with its French garden and geometric shapes. View to the north-northwest from the Gloriette towards the palace.]

In comparison to the Christian relation to the genius loci, it is a matter of an even more radicalised relation of superstructuring what we see here. In order to discriminate it from the Christian relation of superstructuring, as it is qualitatively different, I call it the *relation of excessive planning (overplanning)*. There are two reasons for it: On the one hand it reduces the found place, the found and naturally grown landscape to a purely abstract plane surface, a grid square, where all the mentally-logical, architecturally-geometrical coordinates can be applied; on the other hand the encountered reality has to be literally levelled, i.e. flattened, planed in order to put the plane surface into reality; i.e. to destroy the initial singularity.

2.5 *The romanticisation of the genius loci*

It is in the Romantic era when the antique and medieval concept of the genius loci finds its way into art and science; this happens, of course, under romantic auspices, which means fluctuating between the poles of idealization and unconcealment. Especially in the romantic philosophy and aesthetics of Carl Gustav Carus (1789-1869), an adherent of Goethe, a psychological understanding of the antique phenomenon of the genius loci developed, which wrongfully fell into oblivion and not until half a century later was reintroduced in the scientific debate by Ludwig Klages (1872-1956) and Carl Gustav Jung (1875-1961).

[Pic. 18: Caspar David Friedrich "Nebel im Riesengebirge" 1820/21]

The romantic rediscovery of older genius loci concepts did not only take place as a subject of poetry and philosophy but also in the field of landscape gardening, thus having practical consequences, which can be best studied in the history of the English garden. From the beginning the English garden was themed in a double way: *first,* allowing the natural "spirit of the place" to get a chance when designing the garden; *second*, daring to "jump the fence" by integrating the surrounding landscape and its character. Both ways are expressed in Alexander Pope's (1688-1744) epistle to Lord Burlington (1731): "In all, let *Nature* never be forgot. / Consult the *Genius* of the *Place* in all". This is Alexander Pope's advice to landscape gardening humans: Never forget nature, be guided by local and landscaped realities and strive to allow the natural forces of self-design to get a chance. If you do so, says Pope, then the genius loci can contribute in a creative way. Landscape gardeners then can work hand in hand with nature: What the gardener plants makes the genius loci grow together to a painting; what the gardener initiates with his work, the genius loci creates and accomplishes.

This advice by Pope was not realised immediately, it took almost one hundred years until this advice – arisen from a pre-romantic sensibility – was put into practice by romantic landscape artists like Friedrich Ludwig von Sckell (1750-1823) and Hermann von Pückler-Muskau (1785-1871). I would like to illustrate the practical consequences of the Romantic return to the genius loci using the example of the English landscape garden in Muskau in Upper Lusatia, which was laid out by Pückler-Muskau between 1815 and 1845. Admittedly the existing landscape is reshaped but not deformed. There is therefore a relationship of structuring here – not a relationship of superstructuring or even overplanning. The landscape is rather highlighted, is emphasised in its character. And it is not contradictory at all that one resorts to patterns of landscape painting, quite the contrary. Even if a heroic, classic or melancholic view is orchestrated the existing landscape is not bypassed or ignored. Old venerable trees may remain or their impression is intensified when shaping the landscape.

[Pic. 19: English garden in Muskau.
View from the Belvedere on Lucknitzer Hill of the town and castle of Muskau.]

2.6. *Phenomenological unconcealments of the genius loci*

After the age of Positivism (under the auspices of the Enlightenment) which relegated the romantic tendencies of unconcealment of the original genius loci concept into oblivion (with catchwords like natural science, technology, progress, evolution), a reconsideration of non-reduced phenomena of life and human existence started at the beginning of the 20th century. In the wake of this reconsideration appeared the phenomenological return of the genius loci.

The following theoretical approaches are relevant:

- The expressive-psychological-phenomenological concept of local appearances/physiognomies by Ludwig Klages (Kozljanič 2004b, 262-300).
- Martin Heidegger's onto-poetic-phenomenological concept of the 'place that gathers being', which was concretised by Christian Norberg-Schulz (Kozljanič 2004b, 301-320).
- The bodily phenomenological concept of "spatially effused atmospheres." by Hermann Schmitz (cf. chapter 4.)

As Heidegger's onto-poetic-phenomenological concept of the "place that gathers being", which is the mostly adopted approach in architectural theory and human geography, I would like to sink in it more closely. Later I will dedicate myself more extensively to the handier and phenomenologically more adequate approach made by Hermann Schmitz. In my opinion, Schmitz's approach has the most content and is the most promising – and also the one which complies best "unbiased life experience". The approach made by Klages, by comparison, is rather devoted towards the past and mytho(phenomeno)logical: and is therefore qualified best to shed light on the archaic-mythical and Olympic-mythical experiencing of place and landscape. I will also come back to the Klagesian approach. It is still indispensable, also concerning his understanding of psychological and phenomenological expression. But let us begin with Heidegger in the first place.

In his 1951 essay "Building Dwelling Thinking" Heidegger exemplified by reference to a bridge what the nature of a building is. According to Heidegger a bridge is a built thing. A thing in a special sense here which gets to the heart of "essence" resp."being", concentrates or, as Heidegger puts it, "gathers".

> "To be sure, the bridge is a thing of its *own* kind, for it gathers the fourfold [earth, sky, divinities, mortals] in *such* a way that it allows a *site* for it. But

> only something *that is itself a location* can make space for a *site*. The location is not already there before the bridge is. Before the bridge stands, there are of course many spots along the stream that can be occupied by something. One of them proves to be a location, and does so *because of the bridge*. Thus the bridge does not first come to a location to stand in it; rather, a location comes into existence only by virtue of the bridge. The bridge is a thing; it gathers the fourfold, but in such a way that it allows a site for the fourfold. By this site are determined the localities and ways by which a space is provided for. [...] *Accordingly, spaces receive their being from locations and not from 'space'.*" (Heidegger 1971, 154)

This means the building is a gathering thing. It constitutes the place contrasting it from the pure spot. Social-constructivist architectural theoreticians referred to this passage in order to put the case for pure architect's planning. By the architect's plan only places and as a result landscapes are constituted as characteristic or better: constructed. Dörte Kuhlmann, for instance, tends to the view that the genius loci, the essence of a place, "primarily does not exist but is rather detected and designed by architecture. Through an artificial intervention from its surroundings the place is individuated thus obtaining its individual character". Indeed, as Kuhlmann says, the genius loci therefore only seems to have a "secondary moment or perhaps even a parasitic character" (Kuhlmann 1998).

Kuhlmann invokes as an example the Swiss architect's Mario Botta house near Riva San Vitale from 1972. Kuhlmann comments hereto:

> With this strategy Botta was capable of unveiling most firmly the specific property of the landscape [...]. The tower-like body, opposed dismissively against the alpine landscape as a sleek ashlar, clearly reveals how steep the topography of the plot is, an impression that is even enhanced through the suspension bridge leading to the entrance. Equally the sharp discrepancy between the plain white façade stands out against the colourful and rugged alpine silhouette, changing depending on the season of the year (Kuhlmann 1998).

The fact that this is largely pure conceptual play and rhetoric that can be combined at will is evident if you imagine the house in the Alps in a North German dune landscape and then slightly modify the quotation accordingly: 'By doing so Botta was capable of unveiling most firmly the specific property of the landscape [...].The tower-like body, opposed dismissively against the dune landscape as a sleek ashlar, clearly reveals how *flat* the topography of the plot is, an impression that is even enhanced through the *ascending* bridge leading to the entrance. Equally the sharp discrepancy between the plain white façade stands out against the colourful and *alternatingly illuminated dune landscape*, changing depending on the season of the year.'

If Kuhlmann's interpretation were to hit the corresponding point, Heidegger's concept of the "place that gathers being", would not be a case of a relation of structuring but of a relation of superstructuring. For only through the relation of superstructuring including human design creates something completely new, a new principle, at this place resp. site. Pure (desk)design architecture starts at the stage of the relation of surperstructuring at the earliest, often at the stage of a relation of excessive planning (overplanning).

A perusal of Heidegger leads to a different picture. It shows that – albeit the existential place, for that is what Heidegger is ultimately concerned with, only comes into being with the thing – it is the naturally given place, above mentioned by Heidegger as "spot", has always existed in advance. This is meant in the phenomenological sense not in the concrete object sense. A glimpse into the dialogue "Zur Erörterung der Gelassenheit" ("Conversation on releasement"): Here Heidegger tries to determine the nature of region, one could also say of landscape. For this type of landscape-region he coined the term of "Gegnet" ("that-which-regions" or "that-which-spatially-encounters"). He wants to express that it is something essential which encounters you if you are ready to get yourself into it. It is stated there:

> Teacher: What are we then to call the relation of that-which-regions to the thing if that-which-regions lets the thing abide in itself?
> Scientist: It determines the thing, as thing.
> Scholar: Therefore, it is best called the determining.
> Scientist: But determining is not making and effecting; nor is it rendering possible in the sense of transcendental …
> Teacher: … but only the determining.
> Scientist: We must first learn to think what determining is …
> Teacher: … by learning to become aware of the nature of thinking …
> Scholar: … that is by waiting upon determining and regioning with (Heidegger 1966, 77.)

The architectural thing, which is the bridge in our example, first becomes a place-gathering thing being determined in advance by the natural spot resp. the that-which-regions (that-which-spatially-encounters). The result is: The primary base of the genius loci is the reflective-contemplative encounter with spot and region in the sense of a integrative relation. Only then does the next relation (relationship of structuring) become possible, precisely because it rests on the relation of integration and is essentially conditioned by it, although it transforms the spot into a place – but without deforming it.

The gathering activity of the region resp. of that-which-regions (or "that-which-spatially-encounters") finds its most distinct expression in Heidergger's first journey to Greece in 1962; namely when visiting Delphi. Here he speaks of "a shimmer of the sacred spread over the place" and asks:

> Where did this come from? Not from the remains of the temples and treasuries and the buildings stretching high up the slopes, but from the grandeur of the region itself. Although the Phaedriads shining with radiance concealed their glow at this time of day, the steep rise of their jagged walls and the gaping of the dark gorge between them surpassed in a way connected with the murky depths of the elongated Pleistos Valley any human building. The region itself opened up under the high sky, whose clear air the eagle, the bird of Zeus, circled, as the temple of this locality. Only by doing so it unconcealed, concealing its secret, the space to the mortals in which they were able to erect their sacred works – prior to everything according to rank and due: the Doric temple of Apollo. (Heidegger 2000b, 242; trans. by R. Falzmann).

2.7 *The character of a place as a sedimented phenomenon*

Based on Heidegger's works, Christian Norberg-Schulz (1926-2000) presented in his book "Genius Loci. Landschaft, Lebensraum, Baukunst" ("Genius Loci. Towards a Phenomenology of Architecture") a Heidegger interpretation aiming for a similar direction: "Therefore the existential objective of building (architecture) is transforming a spot into a place, i.e. to detect in a given environment the potentially existing sense." (Norberg-Schulz 1982, 18; trans. by R. Falzmann).

However, this is not possible without getting involved with the specifics that you will always find or encounter them "on site". You have to get involved with the region, allow yourself to be affected by it (also emotionally) and encounter it. Norberg-Schulz calls the specifics the "character":

> A last and particularly important step is accomplished with the term 'character'. Character is defined by the *way* things *are* entrenching our examination in the concrete phenomena of our mundane lifeworld. Only in this way are we able to embrace the *genius loci* completely – that 'spirit governing this place' and was understood in antiquity as the 'counterpart' with whom a

> human has to come to terms with if he achieves to dwell there. (Norberg-Schulz 1982, 11; trans. by R. Falzmann).

So the human – if he does not want to barely reside, but live decently, i.e. feel at home being grounded – shall encounter the genius loci as the character of a place and correspond with it. However, the 'character of a place' is something having-become historically and/or naturally grown, it is located *"zeitgeist"* and/or grown "spacegeist", but always something being 'sedimented'.

Characteristic historically developed places are sedimented in a double sense: first, because historically relics and traces deposited in the earth; on the other hand, because in the social and cultural memory of the locals and the people handing down history there, the locally related remembrance sedimented. Both types of sediments overlie and mingle in the historically grown phenomenon of the place. Told and remembered stories attach themselves to relics deposited in the ground; sedimented relics tell resp. initiate stories and evoke memories.

Characteristic naturally grown places are in a double sense sedimented as well: first because they show a number of geological and biological sedimentations; a place in an impressive grotto, for instance, discloses older earth- and rock strata; and a massive oak tree aged hundreds of years with its annual rings is a very special phenomenon of sedimentation as well; and on the other hand because not only in the historical development of the earth and its (botanical and animal) creatures, but also in the historical development of humans and their entire corporally-mental and spiritual organism sedimented in many layers: from the older archaic ones to the recent more rational ones. Regarding this we can state: both types of sediments overlie and mingle at the naturally grown phenomenon of the place. Ancient deeply-sedimented places activate archaic deeply-sedimented experiencing; and archaic deeply-sedimented experiencing can itemise archaic deeply-sedimented places.

Based on the genius loci human living and building should be considerate exactly of these sediments. For places with sediments are characteristic places as a rule, simply places of character; and thus places that are fundamentally important for providing a spiritual home and closeness to the earth for humans. Such places donate homeland. And such places of character are always multi-layered due to their sediments; being multi-layered always "includes": having depth; and that always "includes" having an "aura" as well. This is all too willingly forgotten in a discourse arguing, the "aura" of a place is something present materially, something one can somehow grasp 'unidimensionally' without paying attention to something that has grown and has come into being.

The same thing I told with reference to the character of a place can also be said in general about the term of the genius loci: This is something sedimented, too. And because it is, provided it is sedimented, a multi-layered, depth-possessing thing, it is ultimately "auratic".

And this is precisely the point I would like to emphasize: The question whether and to what extent the "auratic" and the "numinous" character of the genius loci will and should be addressed in the future is probably not so important. What is more important is to recognise and to acknowledge the term of the genius loci as being a manifold sediment in theory and practice and to find an appropriate mode of dealing with it. Hence it is more important to understand the term like the phenomenon of the genius loci as something historically sedimented, layered. We should not forget that history is always multi-layered. In this sense, we could say that history is always historey (= multi-layered history). And the term and phenomenon of the genius loci can only be fully understood if we try to understand history at the same time as historey: as the experienceable presence of various layers inside and outside of us.

3.
Genius-Loci-Related Urban Development – By Reference to Ancient Rome

3.1 *Genius loci in natural surroundings*

I will return here to the picture from Herculaneum already mentioned and discussed in chapter 2 (picture 1). It is important to bear in mind my description above:

> A Roman wall-painting from Herculaneum shows in a very graphic way how the spirit of place was conceived and what it signified. In the painting we can see a small circular altar, around which a serpent winds its way upwards. Its tail touches the earth, its head stretches over the top of the altar. The serpent – unambiguously identified by the inscription to its right '*GENIVS HVIVS LOCI MONTIS*' as a manifestation of 'the genius of this part of the mountain' – is eating the offering of food that has been placed on the altar.[1] To the left of the picture a boy is approaching the altar. In his right hand he is holding a branch. On his head he is wearing a garland. He could be a shepherd boy who, ritually dressed, is coming to this part of the mountain to make a sacrifice or offering to the genius for the welfare of his flock. This conjecture is supported by verses from Calpurnius. In spring the shepherd should first sacrifice – "Invoke with salted meal the gods, / The guardian genius of the spot / The Lars and Faunus" – then let the flock go to pasture (Calpurnius Siculus, 5, 26; Scott 1890, 95).

Another evidence conveys a further insight into the cultic practice that was linked to the genius loci sites. In the previous chapter I already pointed to this text passage. It is a revealing passage, which, as far as I can tell, was referred by the human geographer Yi-Fu Tuan explicitly to the genius loci but was not acknowledged in this context by classical scholars. It can be found in in Marcus Porcius Cato's (234-149 BC) work "De agri cultura" (On Farming). Although the expression genius loci does not appear here, this passage demonstrates quite nicely that the cultic idea existed earlier

1 See Birt (1886-1890, cols 1613-25; col. 1624 is a reproduction of the picture from Herculaneum).

already (namely during the republican times) and was later (during the Roman Empire) characterised *expressis verbis* by the term genius loci. Cato writes – as already quoted in the previous chapter:

> A pig is to be sacrificed, and the following prayer uttered: "Whether thou be god or goddess [*Si deus, si dea es*] to whom this grove is dedicated [*quoium illud sacrum est*], as it is thy right to receive a sacrifice of a pig for the thinning of this sacred grove, and to this intent, whether I or one at my bidding do it, may it be rightly done. To this end, in offering this pig to thee I humbly beg that thou wilt be gracious and merciful to me, to my house and household, and to my children. Wilt thou deign to receive this pig which I offer thee to this end." (Cato, *De agricultura*, 139)[2]

We can infer from this passage that according to ancient Roman vision certain divine local spirits were immanent to certain sacred and numinous places. These spirits/gods are respected. If human intervention is necessary at these places, an expiatory sacrifice is made: It is a sign that these places are treated with awe and gentleness. At their places sacrifices are made to these spirits in order to avoid calamities and to assure an orderly, blessed and fertile life for the people who live there. Furthermore from this passage (and the wall painting in Herculaneum) it is obvious that the term of the genius loci is limited to quite narrowly defined "microlocal" places as it were: The clearing here, close to the estate; this site of the mountain, far from the village, on the high-lying summer pasture. I could adduce further examples and evidence of genius loci relevance. Most of them would refer to a genius loci notion that is oriented *primarily* towards an idea of a narrowly limited and thus clearly and vividly characterised local and numinous singular appearance.

Should we try to describe the relation of a mythical human here to the place in a formula, we could probably speak most aptly about a natural-numinous *relation of integration.* The human subordinates himself to the place he encounters in its grown and autochthonous structure. He integrates himself, adapts to the location. Even if creative interventions, structural changes and other human manipulations occur, they remain comparatively subtle. They basically subordinate to the discovered place and accept its established characteristics and structures. They devote themselves to the existing numinous influences and fateful moments. Examples of a natural-numinous relation of classification are: the spring sanctuary in its natural state, where a small niche was carved out for the cult figure, a tree sanctuary provided with a

2 Ash (1934, 121 = §139).

subtle enclosure, a mountain or river sanctuary with a small altar showing the veneration for this site.

Characteristic nature-dominated places tend to maintain their spirit through themselves or through the expressive-atmospheric qualities and properties in situ. Examples of such nature-dominated places are: characteristic springs, hills, creeks, rocks, coves, river places, groves, caves. The nymphs are, of course, important spirits of places too, in antiquity and throughout art history: the often lovely spirits of trees, springs, mountains and bays.

[Pic. 20: Nymphae Querquetulanae.
Tree nymphs from a sacred oak grove near Rome. Silver denarius, 43 BCE.]

Genius loci and locus amoenus

The art-historical topos of the *locus amoenus* traces back to such nymphic places. Famous is the *locus amoenus* described by Longus (2nd century) in his novel "Daphnis and Chloe" with its classical triad of spring, tree, rock. Even the muses have originally been nymphs themselves, singing and dancing mountain nymphs (Greek: *oreades*). Hippocrene, their sacred spring at Mount Helicon, where they appeared to Hesiod (cf. Hes. theog. 1-10), is still the symbol of poetic initiation and inspiration.[3] The place and the

3 Cf. Curtius 2013, 183-202 ("The Ideal Landscape"), 228-246 ("The Muses"), 474f. ("The Poet's Divine Frenzy").

atmosphere of Hippocrene can still be seen and sensed today and have the same classic triad: spring, tree, rock (picture 21). This triad has mythological and iconographic ideal-typical, even archetypical significance.[4]

[Pic. 21: Hippocrene at Mount Helicon.]

Locus amoenus and *genius loci* appear again and again in art history as commonplaces – often with a tendency towards the aforementioned triad. Here are just five examples:[5] 1. The inscription “HVIVS NYMPHA LOCI ...” on the painting “Sleeping Nymph” [pic 4] by Albrecht Dürer (1471-1528) (See: Kurz 1953, 171-177). 2. The painting “Reclining Nymph of the Spring” by Lucas Cranach the elder (1472-1553) with inscription “FONTIS NYMPHA SACRI SOMNVM / NE RVMPE QVIESCO” (“I am the nymph of the sacred spring. / Do not disturb my sleep. I am resting”). 3. The famous demand in Alexander Pope’s (1688-1744) “Epistle to [...] Burlington”, which is significant for the history of the English landscape garden: “Consult the Genius of the Place in all” (Pope 1731, 7; cf. Kozljanič 2004b, 207-238).

4 To the best of my knowledge, the first to point out this decisive triad was the poet Heinrich Heine in his writing “Elementary Spirits” (“Elementargeister“, 1837). Cf. Kozljanič 2004b, 38f.

5 Further examples can be found in: Hunt 2022.

4. The "Serpent Stone" erected in honour of Goethe in 1787 in the English landscape garden on the Ilm (Weimar) with the inscription "Genio huius loci" "To the spirit of this place". 5. The painting "A Naiad or Hylas with a Nymph" (1893) by John William Waterhouse (1849-1917).

[Pic. 22: Reclining spring nymph by Lucas Cranach, 1530-1535.]

3.2. *Genius loci in the city – the example of the Roman Forum*

The examples given so far relate to nature-dominated rural genius loci places. However, it is not only there, but also in the culturally dominated urban area genius loci sites are to be found or numinous places where a local spirit, a local deity resp. a local daimon is adored. In that regard we have to verify whether and how a relation of integration can be stated and what consequences arise from it for a genius-loci-related urban development and planning.

Unlike natural sites, characteristic cultural sites obtain their spirit primarily through social references, human imprints and immaterial inscribings. For example, through cultural codifications and interpretations, historical traditions and stereotypings, collective memories and narratives, social functions and psychic projections and, last but not least, through architectural design and structuring. So while natural genius loci sites largely speak for themselves, at least atmospherically and physiognomically, cultural sites require fuller explanation and/or instruction.

Numerous imperial testimonials provide direct evidence of the domestic significance of the 'genius loci', or to the genius of the home.[6] Many inscriptions bear witness to a village or civic cult: to the genius of a village, a district, the place for the assembly, the food stores or the granary, the customs office, the watch-station, the theatre, the threshing floor, the meat market, the school, or the baths.[7] A genius of the port is known from Ostia among others; or a genius of the city from Lugdunum (Lyon). Even in a military context the genius is not without its significance – as the 'genius exercitus' (cf. Romeo 1997, 603f.).

[Pic. 23: Genius from Lugdunum (Lyon). In front of the genius on the left a sacrificing Roman colony founder. Terracotta medallion, 1st century.]

6 For example, *genio domi suae* [...] *aram* (CIL, vol. 8.1, no. 2597, p. 307); ⟨*Io*⟩*vi o. m. et* ⟨*He*⟩*rculi et* ⟨*Sil*⟩*vano et* ⟨*Ge*⟩*nio* ⟨*do*⟩*mus* (CIL, vol. 13.2, no. 8016, p. 539); cited in Otto (1910, col. 1167)

7 See, for example, the references to *genius vici* (CIL, vol. 8.1, nos 2604 and 6352, pp. 308 and 591) or *pagi* (CIL, vol. 5.1, no. 4909, p. 515); *genius curiae* (CIL, vol. 8.1, no. 1548, p. 189); *genius conservator horreorum* (CIL, vol. 6, no. 236, p. 46) or *tutelae horreorum* (CIL, vol. 2, no. 2991, p. 406); *genius portorii* (CIL, vol. 3.1, nos 751-52, p. 142); *genius area frumentariae augustus* (CIL, vol. 8.1, no. 6339, p. 590); *genius macelli* (CIL, vol. 2, no. 2413, p. 339); *scholae* (CIL, vol. 8.1, no. 2601, p. 308); *thermarum* (CIL, vol. 8.1, no. 8926, p. 761); cited in Birt (1886-1890, cols 1621-22).

Genuine genius loci places have priority: these are places that are defined by their unique and irreplaceable character, a character that does not amount to nothing more than pure functionality, disposability and interchangeability. Arbitrary, transferable, interchangeable 'pseudo-genius-loci sites' are of lower interest. In Roman late antiquity these 'pseudo-places' were increasingly common, especially because of their function, their daily easy to handle practicability, their arbitrarily disposable purpose for humans; places for instance were chiefly understood in a functionalistic way as terminological impersonations and adored as such like the genius of a military forces camp, of a meeting place, of gates, of the theatre, of the meat market, of the educational establishment or the thermal baths etc.

I have already mentioned above two paradigmatic examples for the naturally-numinous rural genius loci sites – the example from Herculaneum and the example of Cato. Hereafter I would like to present some paradigmatic examples of artificially-numinous urban genius loci sites. Hereto I will take a look at the main square of the antique city of Rome: the Roman Forum, which was the centre of public life in Rome over centuries. There are some places that could be qualified as genuinely artificially dominated genus loci sites. These are: first, the Umbilicus Urbis, the umbilicus (navel) of the city; second, the Lapis Niger, the sanctuary beneath the black stone; third, the little temple of Venus Cloacina; fourth, the spring of nymph Iuturna.

[Pic. 24: Forum Romanum]

The Umbilicus Urbis

Originally, there was a marshy valley on the site of the forum. The first prehistoric settlements were located on the surrounding hills. From the 10th to the 8th century BC the marshy lowland merely served as a necropolis. But it was only after the (mytho-historical) city foundation in 753 BC that the dead were buried on the Esquiline Hill. When the Tarquinians had the Cloaca Maxima built, Rome's main drainage dykes, draining the swamp, the Roman Forum came gradually into existence with its meeting places, sanctuaries, basilicas, porticos, memorial statues, triumphal arches. So the Forum evolved to a religious, economic, legal and political centre of the city. Together with the city foundation, according to mythical tradition, correlates immediately the construction of the Umbilicus Urbis, resp. its subterranean part called mundus (cf. Pescarin 1999, 26, 31).[8] As we can read in Plutarch's work (Rom. 1,2) the mundus was a ritually excavated pit laid out by Romulus as the centre for the city to be founded where various offerings were placed. Not much is known about the significance of the mundus. If we regard the word only we may assume from the various ways of translation – jewellery, world/space, sky, earth, humanity – that the mundus as the city centre was equally considered as the point where earth and sky, space and humanity culminate: hence a *fourfold* mythically-cosmic city centre. The sparse clues found in ancient literature confirm this assumption:

> According to Cato (Fest. 154 M.273 L.) Mundus signified a pit that looked like an image of the sky (circular), *ut ex is qui intravere cognoscere potui*. The bottom was consecrated to the Manes and therefore closed during the whole year except for three days 24th August, 5th October, 8th November. On these days *mundus patet* (Ziegler 1979, col. 1463).

So the mundus is open on three days offering direct access to the underworld, to the world of the manes. During these days of commemoration of the dead activities, like battle, recruitment or marriage, were forbidden. Above the mundus elevates the small and inconspicuous – probably originating from the 2nd century AD (Groß 1979, 605) – and even today still visible domed structure of the Umbilicus featuring a small door: some sort of temporary gate to the

8 Cf. on this, but also on all subsequent genius loci sites located at the Forum Romanum: Rieche (1978, 120-124, 306f.)

underworld. The word Umbilicus as such, meaning navel, refers to a central, original force or quality that lifts out of darkness into the light, that gives birth and brings forth something.

So this building can be addressed as a genius loci site, not least because it is – through the connection to the subterraneans (Inferi) – a daimonic-numinous place, because it is inhabited by daimons (manes) and because there is a direct connection to a well defined locality, a connection to a site that seems not to be replaceable or transferable. Hence not transferable and replaceable, because what is so (mytho-) historically connected with this site – the foundation and sacred orientation of the city by Romulus – makes this place so unique. By the way, this uniqueness can still by surmised today. Although the knowledge of this place and its former ancient significance (conveyed through archaeology and ancient philology), which the modern viewer usually carries with them as cognitive and literary luggage, is likely to contribute significantly to this experience of the uniqueness of the place, the archaic character of the building – an archaic character that can otherwise only be experienced today at Lapis Niger – is also striking regardless of such bourgeois educational prerequisites. Is the aura surrounding the Umbilicus Urbis not ultimately founded therein? It is difficult to assess. In any case the (cognitive and external) place seems to have an intrinsic activity and emanation, which makes it quite reasonable to allot to it a "Werdegeist" ("nascent spirit", "spirit of becoming") – a genius loci.

When asking about the underlying human-place-relation, it is remarkable that it is not a matter of relation of integration primarily. The reason for this is that the drainage of the marshy valley, which was directly connected with the creation of the city, seriously changed the character of the place. Humans did not subordinate to the marshland. They placed their interests – in this case: intra-urban gain of space through drainage by utilising the marshland – over the natural place they had found. Thus we cannot speak of a relation of integration, but of a relation of structuring. *The place humans found was structured and transformed. However, transformation does not mean deformation here. On the contrary: the "basic form" they found is transformed by an encompassing form; but in a way that through the "trans-form" the character of the "basic form" can still be expressed.* Finally the choice of the place, despite the urban usurpation of the marshy area through drainage and development, gives evidence of an attentive and appreciative transformation of existing circumstances. Caves, ravines, wells, murky lakes and finally also marshes are considered to be entrances into the underworld in the realm of mytho-polytheistic religiosity. At all these places the mythical experience of leading down and/or coming up from the bottom is possible. In such places, the underworld is very close, present and phenomenally tangible for mythical people. So if they placed the Mundus, the connection to the underworld, at the fringe of a drained marsh – where the marshy plain and the foot of the Capitoline Hill converge – this fact speaks for a transformative but not deformative approach to the place. This was all the more the case as the marshland had already been used as a necropolis by the time the town was founded and was therefore designated as an "underground site ". In other words, a sensitive urban planning response was made to both the existing cultural and natural significance of the site. The site was structured – within the framework of urban planning necessities. In other words, the numinous-underground character was indeed reshaped by draining and building. Natural marshland was transformed into an urban area. *But the "citification" of this area did not eradicate or deform the chthonic, numinous-underground character of the site. An artefact, the umbilicus, was used to reshape the original location, but without destroying the underlying character.*

This reshaping preservation of a site, i.e. the relation of structuring has to be realised in its peculiarity. It points out a certain difference in comparison to the relation of classification that existed at the natural-numinous site. The human way of aligning oneself there was not an act of structuring, reshaping, producing. It was an act of sparing, minimally influencing and thus preserving and accentuating (without any extensive structural change and reshaping). Whereas in the relation of structuring, preservation is only possible in passing through extensive reshaping measures. However, the risk of deformation - and thus a relationship of superstructure – can never be completely ruled out.

A lot more could be mentioned concerning the integration-relation, structuring-relation and superstructuring-relation. I did it as stated elsewhere (Kozljanič 2004a, 245-263, 375-401).

But here it is enough to point out that structuring/reshaping can involve extensive structural interventions and yet neither deforms nor superstructures the existing site and its 'basic form'. The discovered 'basic form' is reshaped by an encompassing form; in a way that in view of 'reshaping' the character of the 'basic form' (connection to the underworld, the character of the necropolis) can be expressed.

[Pic. 25: Umbilicus Urbis]

The Lapis Niger

The peculiarity of the relation of structuring becomes clearly apparent in the second example of a genius loci site: the Lapis Niger. It is a rectangle on ground level laid out with black marble plates. Like the Umbilicus Urbis the Lapis Niger is an archaic relic from the origins of the city:

> Although most of the monuments were moved or removed over the centuries, some characteristic buildings survived the test of time almost unscathed. Among them is the sacred precinct with the Lapis Niger, the remains of which still lie opposite the Arch of Severus. When the square was repaved with travertine in the 1st century BC, the rectangular area was covered with black marble plates, presumably to define its boundaries. The remains of a small sanctuary were found beneath it: a tuff altar, part of a column on which presumably stood the cult figure, and a stele from the 6th century BC with an inscription in archaic Latin. The latter, presumably a 'lex sacra', contained a condemnation formula directed against anyone who desecrated the sacred place. The place was also considered sacred, because according to tradition Romulus and Titus Tatius came to an agreement after the battle between the Romans and the Sabines, and Romulus is said to have died there. (Pescarin 1999, 30.)

Plutarch handed down that Romulus died in the middle of the Forum and that a sanctuary was erected above his tomb (Rom. 27, 6) – just the one that can be found today beneath the Lapis Niger.

There is a structuring relation to the discovered site both in the case of the Umbilicus Urbis and the Lapis Niger: also the latter – in the proximity of the Umbilicus, barely forty meters apart from it – underlines the subterranean manes-like character of the former marshy area; it is also most closely connected with the (mythical) history and the fate of the town. Tradition has it that the tomb of Romulus is situated there, the sanctuary of the Lapis Niger obtains the character of a heroon, the sanctuary of a hero and the quality of a daimonic place indeed. Romulus is, so to speak, immortalised there as a protective spirit of the entire city.[9] Not least this site is of much more characteristic and "personal" importance as a "microlocus" than the Umbilicus.

When we imagine – and especially when approaching this place *in situ* – we feel urged to ask ourselves: Does the genius (patris) of Rome inhabit this place? The carved inscription in ancient Latin – which corresponds only rudimentarily to classical Latin can only be roughly interpreted – points in the same direction: A curse upon those who dare to desecrate this central, numinous site.[10]

A further difference is shown here quite clearly between the urban and rural genius loci sites: the urban ones are essentially determined by historical-fateful processes, the rural ones essentially by natural-fateful processes. *The evolving process shapes the site in both cases: Solely the first one is determined first and foremost by fateful-temporal moments, which marked the site, the second one first and foremost by natural and spatial moments, which shaped the place in an unmistakable, numinous way.* I am saying "first and foremost", because in the urban genius loci site the natural and spatial development is not without importance (in the case of the Umbilicus Urbis and the Lapis Niger even very important!), and in the case of the rural genius loci the fateful-temporal aspect can also play a more or less important role (just when thinking about the mytho-fateful incident of Trophonius' oracular cave or Apollo's conquest of the Python or similar local mytho-historical incidents connected with genius loci places dominated by nature). And yet: the influential agent in the urban area is

9 "Individual outstanding personalities were honoured with heroic awards after their death. City founders were particularly often heroised. However, all heroes were regarded as people who had really lived and possessed a special power beyond death." Geisau (1979, col. 1104).

10 Today the original stone is in the museum, only a replica on site.

the fateful-temporal incident, which buries itself in the site and confers to it its distinctive character. After all the Umbilicus is not *just any* access to the underworld, not *any* central point *of some sort*, but *the* site where Romulus founded Rome, *the* site which is most closely linked with the history of Rome's origins, where this history of origins is more palpable than *anywhere* else. And the like pertains to Lapis Niger: It is not *any* stone *of some sort* either, but *the* site where most important historical incidents took place (agreement with the Sabines; death and tomb of Romulus), incidents of fateful conciseness and implication. And *this* is the central aspect, this is the fact that makes this site ultimately artificially-numinous, a fatefully-numinous and unique site.

This means that there are not only relations of structuring in the artificial-urban area but in addition also relations of integration. Therefore a double relation: in relation to the naturally-geographical site (the subterranean-manes-like marshland) we deal with a relation of structuring on the one hand, on the other hand in relation to the fateful-temporal site (foundation of the city and death place of Romulus) we deal with a relation of integration instead. Integration of the urban intervention means, above all, integration in what is fatefully and numinously connected with a place, what has become engraved there mythologically and historically. In the case of the Umbilicus all that embedded itself historically in the site was preserved – in concrete terms: the mundus that was trenched by Romulus; the only interventions which occurred in succession were very subtle. They protected and accentuated this site by the small domed structure of the Umbilicus, not more. So it is not a structurally extensive act of reshaping, but a retaining, protective, revering and preserving one. This shows that the urban way of dealing with history-charged, fateful, artificial genius loci sites – unlike the urban handling of natural places embraced by urban expansion – resembles the rural way of dealing with nature-dominated genius loci places.

The Venus Cloacina sanctuary

The peculiarity of the urban human-place-relation can be shown in the third example as well, the Venus Cloacina sanctuary. This sanctuary allegedly founded by Titus Tatius is only about 80 meters away from the Umbilicus, and barely 50 meters away from the Lapis Niger, hence in the central position of the Forum in immediate proximity of the forum square and the Curia. Today we can just see the circular form of the foundation of this small sanctuary. It once was an open building housing two images of goddesses, which are represented by the L. Mussidius Longus (39 BC) denarii. The small sanctuary was at the point where Rome's main drainage canal, the Cloaca Maxima, reached the forum (Eisenhut 1979a, col. 1226).

> Inside the wall ring there was an altar and two statues of the deity in her two representations as Venus and Cloacina. The name was derived from the Latin word cluere (purify) and referred to the purifying power of the myrtle branch. [...] According to legend the Romans and the Sabine purified themselves here with myrtle branches before fighting each other after the Rape of the Sabine Women. [...] This was the dividing line between the two enemy peoples. (Pescarin 1999, 29)

Although there is no consensus in academic research how to interpret the combination of naming of Venus Cloacina, the general tendency is to see in Cloacina an initially independent local daimon, which was later reshaped by the goddess Venus and finally lived on with her surname only (cf. Eisenhut 1979a, col. 1226). If this assumption were true, we would speak about a genuine genius loci.

This example also shows the existence of an underlying twofold urban human-site-relation. On the one hand there is a structuring nature relation: for this genius loci site also has a distinct relation to the earlier marshland; it does not deny it, but takes up essential issues from it in its new expressive substance. On the other hand there is the integrating, preserving, respectful historical reference: historically fateful incidents like the conflict with the Sabines, the reminiscence of Romulus resp. his rival Titus Tatius, the draining of the marshland by means of the Cloaca Maxima are reverently remembered here; the structurally urban intervention – the building of the Venus Cloacina sanctuary – integrates to these incidents and its manifestations *in situ*.

The Spring of Iuturna (Lacus Iuturnae)

Now, let us have a look at the fourth example, the spring of Iuturna. This spring sanctuary is located at the Southern corner of the Roman Forum, directly next to the Castor and Pollux temple. Before the completion of the first aqueduct in 312 BC this spring, whose water was considered to have healing powers (Varro 1.5,71) and was later still used for all public sacrifices (Serv. Aen. 12, 139), supplied the Romans with drinking water. In republican times already (2nd century BC), the spring water was contained in a large basin with a group of figures. This basin, called Lacus Iuturnae, which changed its appearance under Trajan for the last time, is still visible, as well as a chapel ten metres south-east of it along with the Ara sacrificale. The reconstructed Aedicula – bordered by two columns, closed by a marble tympanon – originates from the 2nd century AD. In front of it there is a round altar bearing this inscription "M BARBATIVS POLLIO / A ED CVR / IVTVRNAI SACRVM / TEAL" as well as an angular predella with a relief on it showing Iuturna and her brother Turnus. Her brother Turnus (the former fiancé of later princess Lavinia married by Aeneas), whom she in vain tried to support was defeated and killed by Aeneas (Verg. Aen. 12, 842ff). The shielding and equally farewell gesture showing Iuturna turning to her armour-clad and armed brother on the relief reminds of this mytho-historical incident. The spring nymph Iuturna is regarded as the daughter of the river god Volturnus, spouse of Janus, mother of the spring god Fontus/ Fons (Arnob. 3, 29).[11] In an inscription found near the spring she is called *Genius Stationis Aquarum* (cf. Schilling 1978, col. 68).

[Pic. 26: Altar of the spring nymph Iuturna, Forum Romanum.]

11 Cf. Pescarin (1999, 36f.) and Eisenhut (1979b, col. 25f.).

The example of the spring nymph Iuturna is very instructive. It shows from another point of view what happens with a genuine once natural genius loci place, as soon as it gets under the immediate influence of urban design and development: it is extensively reshaped and structured (almost superstructured). In fact quite substantially in the case of the Iuturna spring. Not a single square centimetre of the spring sanctuary was left untouched in its natural state; there are walls, pools and channels everywhere. Everything that was a natural spring once is completely reshaped by urban structural intervention. Here is the artificially contained spring water exit, there the artificially contained catch basin; the water discharge is equally artificially canalised underground. And yet: The spring site itself is not deformed, at least as far as the authenticity of the place and its mythical importance is concerned. Not only the urban relation of structuring, but also the urban relation of integration – the classifying reference to the mytho-historical fateful place – can be perceived at the spring of Iuturna: the memory of Turnus' decisive battle with Aeneas, the Trojan ancestor of the Romans. Also the altar base with its Dioscuri relief, which is in the middle of the Iuturna basin (today), points to a mytho-historical incident. After the Battle of Lake Regillus against the Latin League the Dioscuri (Castor and Pollux) were allegedly seen watering their horses at the Spring of Iuturna. (Ov. fast. 1, 706; Dion Hal. 6, 13).

3.3 *The Imperial Fora*

So much for the examples relating to the genius loci of the urban human-site-relation. In order to show that in urban dealings with local realities also completely different human-site-relationships can occur – human-site-relationships that have nothing in common with a genius-loci-relation of integration or a relation of structuring – I would like to cast a glance at the so-called Imperial Fora. The Imperial Fora, as far as the present topic of the genius loci is concerned, can be viewed as the pure opposite of the Roman Forum. If we found many genuine genius loci sites at the Roman Forum, we cannot (almost) find any at all. During the planning and building of the Fora local spirits did not matter, showing consideration for what already existed and came into existence was not the main concern.

What mattered here was the demonstration of power. A look at the development history of these Fora shows it quite clearly. They came into existence from 50 BC until 115 AD in rapid succession. In each of these four Fora a Roman emperor wanted to immortalise himself. Whereas we have to add that the phrase "an emperor in each Forum" is not quite exact. For the first so-called "emperor" – Gaius Julius Caesar (100-44 BC) – only was a Roman dictator officially from 48-44 BC. The first Roman emperor – with whom the republican era ended – was since 27 BC Caesar Octavian Augustus (63 BC - 14 AD). The first new Forum was, as mentioned earlier, Forum of Caesar, the so-called Forum Iulium. It connected immediately to the northern part of the Roman Forum. Connecting eastward the Forum of Augustus that was commenced in 42 BC already but completed in 2 BC only. In direct southern vicinity there is the narrow Forum of Nerva commenced by emperor Domitian (51-96 AD) and completed by emperor Nerva (30-98 AD) in 97 AD. North of the Forum of Augustus we find the biggest of the Imperial Fora: the Forum of Trajan erected between 107 and 113 AD by emperor Trajan (53-117 AD).

The structure of all Imperial Fora is similar although they vary in size and changing orientation (north-south or east-west). There is always a large open area (with one exception)[12] with the dominant statue of the current emperor. In the Forum of Caesar as well as in the Forum of Trajan there is an equestrian statue of the emperor and in the Forum of Augustus the statue shows the emperor in a triumphal quadriga. In each forum there is only one dominant sanctuary closing each time the forum at its gable end but never dedicated to a local god, however, always to a god closely related to the ruling power of the emperor. The temple at the Forum of Caesar was dedicated to Venus Genetrix, mother of Aeneas and mythical ancestress of the Julian lineage, to which Caesar belonged. The temple does not owe its origin to a local myth but Caesar's vow before the Battle of Pharsalus against Pompey. Caesar vowed to build this temple in case of his victory. An episode mentioned by Sueton in his Caesar biography shows that he ultimately had this temple built for himself: One day Caesar received the senate, enthroned on an armchair like a god in front of the temple in the middle of the podium (Suet. Caes. 78).

12 This exception is the smallest and most modest of the imperial forums, the Nerva Forum. As far as I can see, the Nerva Forum is also an exception because there was another sanctuary on it - a sanctuary that did not serve the imperial cult. It was the sanctuary of Ianus Quadrifrons, the four-faced Ianus. The only place in the area of the imperial forums that could possibly be recognised as a genius loci site. Perhaps the place was an important crossroads in earlier times? After all, the so-called Argiletum – a connecting road to the Basilica Aemilia at the Forum Romanum – passed through the Nerva Forum, which was therefore also called the Forum Transitorium. (Hanslik 1979, col. 76).

Caesar thus places himself on the same level as the gods. The self-glorification and self-deification was made a tradition by the following emperors. Then Hadrian brought this tradition to a culminating point: He had a temple built at the end of the Forum of Trajan, which was dedicated to Divus Traianus pater and his spouse Plotina. The temple of the Forum of Augustus was the Temple of Mars-Ultor. Within the temple stood the cultic statues of the god Mars in his guise as avenger and Venus as the ancestress of the Julian lineage to which Augustus belonged. The Forum of Nerva closed with the temple dedicated to Minerva, the goddess of arts and warfare and the patroness of Domitian (cf. Pescarin 1999, 46-57).

The uniformity of the Imperial Fora is striking. Sofia Pescarin indicated that this uniformity was primarily politically motivated and backed by a coherent power-political programme.

> The Roman Forum, whose structure was directly related to the political system of the Republic, was always the scene of heated discussions and parliamentary disputes in the Senate and the Comitia. The Imperial Fora... on the other hand reflect the change of state leadership, which was now in the hands of the Emperor and his confidants. The political struggles took place elsewhere, and the fora increasingly took the task of praising the person of the Emperor who had built them. Architects constructed the buildings precisely according to plan, as they were intended to reflect the ideology of the imperial era, portray the ruler as a powerful generous emperor, and make his government programme visible to everyone. In addition to the court buildings, which were still numerous, there were now an increased number of educational and cultural institutions as well as trading centres. Examples of this were Greek and Latin libraries, Trajan's Column, and the Trajan Market on the forum of the same name. (Pescarin 1999, 47)

If one now asks about the specific human-place-relationship that comes into play here, it is obvious that this cannot be neither a relation of classification nor a relation of transformation. It is not the human being who has to subordinate himself to the place and his fate but rather the place has to subordinate to the ruler's plans of self-glorification and power interests. The place is subjected to a purposively rational human function. This subordination tolerates no relation of transformation. What is found, grown or developed at the site – regardless of whether it is natural or fatefully-historical – is peripheral, irrelevant. The architectural design, which is intended to pay monumental homage to the respective ruler, is applied to the site without regard for losses. One does not accommodate the old one does not reshape it. Old things are destroyed and levelled. This is most evident in the case of the biggest Imperial Forum, the Forum of Trajan.

> The famous architect Apollodorus of Damascus bore responsibility for this project. He also carried out large cut through the saddle between the Capitol and the Qurinal, which created space for a new forum in the densely built-up area. Trajans Column still recalls the original height of the saddle. (Pescarin 1999, 52)

Trajan's Column – decorated all over with relief scenes glorifying the deeds of the emperor, whose ashes were interred beneath the column – is almost 40 meters high.

[Pic. 27: River god on Trajan's Column in the Imperial Forums]

I would like to call this specific, functionalist human-place-relation the relation of excessive planning (overplanning). As I said in section 2.4, this relationship of excessive planning is a radicalised superstructure relationship. I speak of excessive planning because, first, the prior rational planning, which is to be imposed on the place later dominates; this is done in such a way that, second, the place is given no chance to exert any influence on planning and design; the place is sacrificed to other interests, is completely superstructured; third, to ensure this overplanning the place has to be flattened first, must be levelled, like a plane surface in the technical sense (cf. Kozljanič 2004b, 112-126).

A numinous and multi-layered place has no chance of survival under these conditions. A genuine genius loci place – no matter whether it arose naturally in space or fatefully in time – cannot survive under such conditions, nor can it form anew. It disappears making way for the ideologically functionalized place and the power interests objectified with it. Incidentally this can still be experienced quite clearly on site today. For the Imperial Fora still have no charisma, no atmosphere. They are comparatively characterless, dead. No "nascent spirit" has manifested itself there to this day. Quite the opposite: When the current Via dei Fori Imperiali was built by Rome's fascist rulers in the 1930s to create a space for military parades, the area of the Imperial Fora was cut in two. A blunder that had detrimental consequences not only visually but, above all, archaeologically: a good part of the forums was thereby tarmacked. The irony of fate: the Imperial Fora finally fell victim to the same "spirit" that created them: the will to power, the "veni, vidi, vici".

3.4. *Conclusions for future genius-loci-related urban planning*

Before I formulate conclusions for future genius-loci-related urban planning in the context of the insights gained so far, a preliminary remark: Of course, genius loci no longer means local god or local daimon *today*. The mytho-religious genius loci concept of antiquity is (normally) no longer binding or credible for us: Today's visitors to the Roman Forum are *not* of the opinion that a nymph lives at the Iuturna spring, to whom one should offer sacrifices, to whom one can pray, and who occasionally appears in human form. Thus, when human geographers such as Yi-Fu Tuan (1976), architectural theorists such as Christian Norberg-Schulz (1980/1982), urban planners such as Harry L. Garnham (1985), a landscape historians such as John Dixon Hunt (2022) or philosophers such as Gernot Böhme (2017a, 178f.) and Tonino Griffero (2020, 137-149) speak of genius loci today, they mean something different: they primarily mean the quality and atmosphere, the character and peculiarity, the conciseness and expressiveness of a place. Even more: They believe that atmospheric, expressive, and meaningful places are an important prerequisite for quality of life and living, an important prerequisite for feeling comfortable and at home in an area, for being able to become familiar with and identify with it, its history, its residents, and its architectural features. Admittedly, at least some of them also mean something that could be described as the numinous charisma and aura of a place (more on this in chapter 5). I don't want to conceal that, because it ultimately belongs to the modern understanding of the genius loci; because it is somehow subliminally, and in its subliminality almost ineradicably, part of it. What Mircea Eliade once said with regard to nature can, with certain restrictions and modifications, also be said with regard to the city:

> The experience of a radically desacralised nature is a new discovery and, moreover, accessible only to a minority of modern society, primarily to scientists. For the rest of humanity, nature still has 'magic', 'mystery', and 'majesty' and from this one can read traces of the old religious values. (Eliade 1957, 89)

I, too, feel committed to this contemporary understanding of the genius loci. Therefore when I speak of the genius-loci-related urban planning in the following, I mean:

- First, urban planning that takes existing atmospheric, expressive, and meaningful features into account and strives for a careful, gentle, preserving and lovingly compassionate approach to them.
- Second, urban planning that also takes into account *naturally*-local and *naturally*-surrounding landscape features and particularities; for in every city there are such features and particularities, no matter how much they are pushed to the margins, artificially surrounded, reshaped or concealed: rivers, streams, trees, 'natural' rocks, fallow land reclaimed by wild vegetation, etc. (cf. Röhnert 2025).
- Third, urban planning that is besides responsive of *historically*-local *historically*-surrounding landscape conditions and particularities.
- Fourth, urban planning that consciously and purposefully acknowledges open spaces whether in the form of squares, backyards, wastelands (not yet dominated by overplanning); or in the form of ruined residential and factory buildings; or in the form of parks and gardens where there are still niches for wild animals, weeds and other uncivilized beings. Urban planning must provide these open spaces and protect them against all kinds of economic, political and private interest groups, because authentic genius loci places cannot be planned: they become. Above all, however, where open spaces are permitted, new genius loci sites can emerge and grow; it is precisely there that a new 'nascent spirit' can form and materialise. If we not only want to preserve old genius loci sites but also enable new ones, if we not only want to be backward-looking but also forward-looking and enable progressive genius loci-based urban planning, it will be good to pay more attention to such open spaces.

- Fifth, urban planning that, not at least, also takes into account the local residents and their lifeworld perspective (especially and primarily the local people, but also the local animals and plants) integrating them into the design and planning process.

The Hafeninsel city park in Saarbrücken and (although this is more than just urban planning) the 'Meadowcreek' project are proof that these five points can be realised if you want to (cf. Reisinger 1991; Coates, Seamon 1993; summarising and explicating: Kozljanič 2004b, 384-426)

If one now asks what modern, genius-loci-related urban planning can learn from ancient Rome, the following points can be noted:

- One: There are genius loci sites that arose naturally *and* those that arose through fateful history. Although the boundaries between the two are fluid, and although there are numerous transitional phenomena, it can be said that the latter dominate in the urban sector and are of high planning relevance there.
- Two: Naturally arisen genius loci places can be perceived relatively directly in terms of their natural local character (= spatial atmosphere + physiognomic appearance). Their continuity over time need not be created or guaranteed by humans: It maintains itself as long as humans leave it alone and do not disturb or destroy it. Genius loci sites that evolved through fateful history are *different*: They require social and cultural memory, passing on tradition and mediation in order to be perceived in their particularity and their local character. Their continuity over time is guaranteed by human tradition, care, and activity and is therefore something humanly created and maintained. This gives rise to the task of remembering and maintaining genius loci sites. Genius-loci-related urban planning thus means more than mere planning: it also means protecting culture and local history.
- Three: The fact that naturally occurring genius loci places can be perceived more or less directly in their particularity does not mean that their local character is factually and unambiguously given, does not mean that they could be perceived by means of single-sensory, rationally objectifying and judging ("diacritic") perception of mere things. One can only experience and bodily sense the local character by means of holistically sensory quasi physiognomic ("coenaethetic") perception of expressions and synaesthetic characters.[13] In other words: Just as genius loci sites

13 For the depth-psychologically and empirically based concept of "diacritical" and "coenaesthetic" perception, see: Spitz (1972).

became fatefully-historical require "internal remembrance", so too do naturally arisen genius loci places require "internalisation". This means, however: Neither naturally arisen genius loci places nor those arisen fatefully-historical ones are "objective" places. This does not mean, however, that they are "subjective" places. They are rather life-world places: and as such they are subject-object preceding "experienced places".[14] Experienced places "exist" in and with the life-world of people living with them. Urban planning that neglects this fact in its planning not only ignores it, as far as the genius loci is concerned, but also and especially concerning the people who live and work there; concerning the people who perform their daily/weekly/monthly 'place-ballets'[15] there.

- Four: Urban genius loci sites are not only historical but also layered places. They are layered places in two ways: firstly, historical relics deposited and sedimented: rubble, waste, everyday objects, bones; secondly, memories accumulated in the social memory of the residents and/or those who transmit history (in written or oral form). Both forms of sedimentation overlap and mingle in the life-world phenomenon of the site (cf. Freyer 1990). Told and remembered stories attach themselves to relics sedimented in the ground; sedimented relics themselves tell or initiate stories, evoke memories. Genius-loci-related urban planning should take these sediments into account: for only sites with sediments can be multi-layered; but being multi-layered means: having rich depth; having an "aura"; having an aura means: evoking the numinous.)
- Five: The urban character (= bodily felt atmosphere + physiognomic appearance) condenses in certain crosstown city sectors (quarters), centring and anchoring itself selectively in specific places and their objects. Genius-loci-related urban planning should handle such *areas of concentration* and *anchoring points* with great sensitivity. The following applies:
- Six: Pure relations of excessive planning (overplanning) should be avoided in these areas and at these locations, for they destroy atmospheric, expressive and meaningful places. Relations of excessive planning should be limited to such regions that do not

14 Cf. Bollnows „concept of experienced space" (2020, 37-41).

15 The metaphorically ingenious term "place-ballet" was coined by David Seamon. See Seamon 1980 and Eck, Pijpers 2016.

feature any distinctive places: hence to atmospherically "de-densified" regions with little expression and meaning.

- Seven: Relations of structuring should be preferred in densified areas, because they *enable* genius-loci-related urban planning; they should also be considered as a "second choice" when relations of integration are not possible or sensible (due to planning necessities, for example).
- Eight: Relations of integration should be preferred with regard to anchoring points, as they guarantee the greatest possible conservation and continuity of the genius loci.[16]

16 More detailed explanations on the topic of genius loci and genius-loci-related urban and open space development can be found in: Kozljanič 2004b, 345-430.

4.
The Felt-Body-Phenomenological Concept of "Spatially Effused Atmospheres" of Hermann Schmitz

Hermann Schmitz' (1928-2021) is the founder of a "New Phenomenology". One of the key notions of his New Phenomenology is – beside "felt-body", "(own) bodily feeling or condition", "affective involvement", "holistic situations", "quasi-things" – the notion of "atmospheres". He defines "atmospheres" as "spatially extended, each time outpoured without boundaries, bodily perceptible emotions" (Schmitz 1995, 6f., 196-199, 292-296). For Schmitz "*atmospheres*, among them weather and silence (as experienced, not as constructed), [… are] mainly [...] emotions that have to be regarded as spatially effused powers able to affect the body" (Schmitz 2002, 492). Elsewhere, Schmitz says: *"feelings are spatially effused atmospheres and forces that grasp us felt-bodily"* (Schmitz 2023, 45).

There is an excellent brief outline of Schmitz's concept of atmosphere by Tonino Griffero. It is worth quoting it in full:

> In a nutshell, atmospheres are centripetal and external vectors; they have authority and, generating affective involvement (an embodied experience), they grip the vital drive, suddenly taking over and leaving the perceiver based on their specific resonance in their personal embodied disposition. For the ancients, rage, *eros* and parrhesia – for instance – were not private inner emotions but atmospheres, partially personified respectively by Ares, Aphrodite and the Holy Ghost. Also, they were taken to be so external as to be impervious to the human projective impulse, while today such impulse is regarded as constructivistically omnipotent, leading to the erroneous belief that the world is semantically and affectively empty. Be they swift or gradual, atmospheres govern our thymic life. In fact, it is undoubtedly true that we can have the mere perception of an atmosphere without being in the grip of it, and it can even happen that an emotion (as an atmosphere) is in the air without anyone feeling it; however, atmospheres normally grip a person as half-entities (quasi-things or half-things), thus arousing a natural and very precise suggestion of motion, luckily not yet fully repressed by our personal emancipation. (Griffero 2019, 23f.)

Landscape ambiences as it was repeatedly thematised in the course of my research are just such "atmospheres" for Schmitz. Climatic seasonal and diurnal moods, such as the tension filled atmosphere of a thunderstorm, the crystalline mood of a starry, bitter-cold winter night, the muggy sultry mood of a summer midday: All these are atmospheres, or more precisely "super-personal atmospheres". In addition to non-personal, resp. supra-personal atmospheres there are also "person-bound atmospheres" such as joy, sadness, bliss; these "person-bound atmospheres" can seize single but also more people and radiate from them. Schmitz also knows and names "religious" or "divine" atmospheres such as the atmosphere of "Christian love" or numinous locally bound atmospheres (cf. Schmitz 1969, 98-133).

No philosopher has perhaps engaged so intensively, precisely, and in a disciplined manner with atmospheric and mood-related phenomena as Hermann Schmitz. I would like to explicitly emphasise here how much phenomenological and aesthetic discussion of such phenomena, which has stalled for decades, could be enriched and advanced by the Schmitzean concept of atmosphere. For many of today's phenomenologists and aestheticians have not realised Schmitz and his concept of atmosphere. And among those who have engaged with this concept, some have not realised its significance or the theoretical and practical applications it offers. Fortunately, there are exceptions. One exception is Gernot Böhme (1937-2022), who has long pointed out the importance of Schmitz's felt-body-phenomenology and has applied and modified his concept of atmosphere in numerous aesthetic and phenomenological investigations. Not least, Böhme's studies demonstrate how and in what ways the concept of atmosphere proves valuable, and they also reveal the forward-looking potential this notion contains. Böhme's assertion that Schmitz's concept of atmosphere should be one of the central terms in a future "aesthetics of nature" and "phenomenology of nature" is entirely to be supported. Moreover, regarding a broader understanding of human self-concept, Schmitz' concept of atmosphere holds a pioneering potential:

"A rehabilitation of atmospheres would enrich human life with experience and free our understanding of humans from the constriction in which magic and sorcery are dismissed as mere deception, and the world of pagan gods is regarded as null or subjective projection. The bodily reality of humans themselves could also be appreciated differently." (Böhme 1985, 206; see also Böhme 1989, 1995 and 1997).

Today, it is Tonino Griffero who is further developing Schmitz's New Phenomenology and Gernot Böhme's Philosophical-pragmatic Anthropology in a very profound and creative way (see Griffero 2017, 2018, 2019, 2020, 2021). It is also very important to mention Michael Großheim, holder of the Hermann Schmitz Endowed Chair at the University of Rostock. Großheim, his colleagues and the main representatives of the "Society for New

Phenomenology" (Gesellschaft für Neue Phänomenologie) have done crucial work, not least in the series "Neue Phänomenologie" published by Alber Verlag (see: https://www.gnp-online.de/publikationen/gnp-buchreihe.html).

Back to Schmitz and his definition of atmospheres: It is true for all atmospheres that they are initially "spatially extended and borderlessly outpoured". Atmospheres are feelings "that can poignantly settle together over many people; but also such feelings that are considered to be the private matter of the individual, are equally spatially extended." (Schmitz 1995, 293f.). Due to their spatial outpouring, personal-bound atmospheres differ from purely "corporeal impulses" such as hunger, thirst, fright, lust, disgust, freshness, faintness, etc. (Schmitz 1965, 173-253). For corporeal impulses remain closely bound to the body whereas personal-bound atmospheres radiate in a borderless way into the environment from those affected by them.

This may sound unfamiliar. However, Schmitz' distinction between "corporeal impulses" and "personal-bound atmospheres" is something that most people should be familiar with becomes evident in the following example:

> With a certain degree of sensitivity, the sad person in cheerful company and the cheerful person in sad company will have a pronounced experience of contrast. The latter, for example, will feel somewhat uncomfortable and embarrassed, as if his cheerfulness were out of place amidst so much sadness; the former will find the contrast offensive, so that the protest question arises: 'Why must I be so sad amidst so much cheerfulness?' [...] If, again, the dull person finds himself in a fresh environment and the fresh person in a dull one, such a social contrast is far less motivated. The dull person will not give rise to such a fundamental protest as in the previous case. His question: 'Why am I so much duller than the others' will rather arise from reflection on causes and goals and perhaps lead him to the physician. [...] Corporeal impulses are spatially circumscribed [i.e., they have their comparatively clearly defined bodily location ...] and can therefore more or less smoothly exist side by side even as incompatible ones; whereas emotions as borderlessly outpoured atmospheres make a total claim to captivate the entire area of the present stage of events, with everything that takes place on it, and thereby entangle the person seized by them in the embarrassing contrast when their aspiration bounces off the incompatible, represented by a contrary feeling, which assert itself equally or more strongly in the constellation in question. (Schmitz 1995, 295f., cf. 118f.)

So far concerning the personal-bound atmospheres and the corporeal impulses. In what follows, I will concentrate on the various genius-loci-relevant, not personal-bound atmospheres.

4.1 *Climatic, seasonal, and diurnal atmospheres as expansive-spatial feelings "hanging in the air"*

For Schmitz, the outstanding paradigm of non personal-bound, supra-personal, everyday atmospheres are "climatic atmospheres". These include all sorts of climate-like, weather-like atmospheres, including the particular seasonal and diurnal atmospheres. At least in their more obvious manifestations, everyone can feel such atmospheres bodily. When I step out of an air-conditioned room into the sultry midday summer heat of a large valley basin, then I notice how from outside – mind you, and not projected by me – the oppressive, burdensome paralyzing nature of the atmosphere envelopes me and starts to penetrate my body. If I overcome the resulting weariness and climb – through shady forests that somewhat soften the sultriness rendering it more bearable, over the pasture of an alp, where the first vitalising breeze is noticeable – over one of the surrounding peaks, it can happen that up there, far above the valley floor, I am surrounded by a completely different, more refreshing and more vitalising atmosphere, which was not "brought along" by me either, I found it up there. These atmospheres are admittedly bodily noticeable in or on my own person. However, they are not personal-bound, but "supra-personal"; or should I better say: 'extra-personal and sub-personal'?

Schmitz has demonstrated the peculiarities present here using similar examples and analysed them felt-body-phenomenologically:

> Particularly revealing for the more primitive layers of spatial experience, that matter here, is the [...] climatic sensing. In a room, we immediately notice whether it is humid or stuffy; in the high mountains or high forest we immerse in a fresh, cool, austere atmosphere, which we sometimes painfully miss when returning to the lowlands, where it is humid and balmy. What is the subject of these climatic attributes? We like to help ourselves with impersonal sentence constructions or with metaphors, for example when speaking of air; this speech is metaphorical insofar as the phenomenal climate reveals nothing of a gas, but makes us think of air only in the sense of an atmosphere that is somehow laden, suffused or dull [...] This atmosphere in which we find ourselves embedded thanks to our perception of weather or the interior climate, is each time outpoured into space without boundaries, that is each time involuntarily given to us as the background of our corporeal condition, on this side of all theoretical knowledge and the realisation of all other subjects where the climate [...] would be different. This vastness is an atmosphere of perceived climate without layers or distances, voluminous and indivisibly expansive. [...] This atmosphere is felt in one's own body, but not as a condition of one's own body. The wet and cold November weather, for example, is felt in one's own shivering, in the stuffy

> air, in a dull pressure that weighs on the body and prevents deep breathing; the sultry summer thunderstorms are felt in a nervous, electrifying restlessness in one's corporeal condition; the gentle spring air in a melting physical comfort. In doing so, one feels [...] certainly not the condition of individual parts of the body such as the skin, throat or lungs, but something that 'hangs in the air' as climate, in a sense that the throat, skin and lungs do not hang in the air. (Schmitz 1995, 292f.).

As the climatic, seasonal, and diurnal atmospheres "hang in the air" in this way, they remain only vaguely definable as far as their spatial mode of appearance is concerned. The spatial mode ascribed to them, according to Schmitz, is that of a "wide space" or "expansive space". The expansive space consists of two components: the bodily felt atmosphere and the own sensing body. Directions, areas, positions, distances and relative, i.e. conflictingly definable locations do not exist in the expansive space. According to Schmitz:

"In your own felt body you sense a vaguely outpoured climate, in which you find yourself, and your own body, which, at his absolute location [...] stands out as affected. Further structuring cannot be derived from the mere sensing of the climate." (Schmitz 1995, 280; cf. Schmitz 1969, 361).

[Pic. 28: Atmospheric landscape near Zugspitze]

4.2 *Landscape atmospheres as directional spatial, condensed and object-related emotions*

The situation is different with regard to supra-personal atmospheres that are not actually "in the air" but rather "in the landscape". Such atmospheres are less ephemeral, they are rather "objective" resp. "matter-related". They are related to "objects" (for example to "objects" of a landscape, such as hills, plants and villages) and are present there. Consequently, they acquire, beside their "spatially extended" definiteness, a "spatially directional" one as well. Schmitz defines the "directional space" as follows: "In the corporeal directional space, directions arise between the narrowness of the felt body, from which the absolute location [...] emerges as its spatial aspect, and the wideness that lead from the narrowness into the expanse, and emerge from these regions, but do not yet find a definitive destination in relative [i.e., clear, distinguishable and definable] locations." (Schmitz 1995, 281). Schmitz refers to such directional atmospheres as "object-bound" or, briefly, "objective emotions." (Schmitz 1969, 98, 367-374).[1] *However, I prefer to speak not of "objective" emotions, but rather of object-related emotions/atmospheres in the context of landscapes, and of object-bound emotions/atmospheres when referring to locations* (see the next section). The terms "objective" and "object-bound" risk quickly falling into the trap of a worldview rooted in a split between subject and object, a rational and reifying perspective. Yet, such a worldview is either absent or only present in a nascent or extreme form within the realm of spontaneous and naive "corporeal sensing."

Regarding the "object-related" and "object-bound atmospheres," Schmitz states that they "stand out, like a nimbus or halo, around certain objects, for example, optically given ones, and [can] be subjected to the same categories of spatial arrangement as these objects themselves." As an example, he mentions, among others, a seascape to be cheerful and emphasises that we must not err by imagining this landscape as a person capable of feelings. For: "The sea is only cheerful in the sense that a corner where it roars can be loud without being heard; its cheerfulness is an objective emotion," and not a subjective one, i.e. felt by a human subject.

1 Landscape atmospheres are, of course, only one of several types of "objective emotions".

Following this and for the sake of confirmation, Schmitz now quotes Otto Baensch. Due to his, as Schmitz also emphasises, "exemplary clarity," it is reproduced here once more:

> Besides subjective feelings, which move our inner selves, or those that reveal themselves to us in the bodily expression of other animated beings as their inner states, the world is filled with objective emotions. These emotions adhere to objects, surround them, and penetrate them, without needing to be regarded as states of a soul residing within the objects. [...] We do not think of the landscape, however blurred, as a sentient being whose outer form expresses an inner mood as an 'expression' of something internally present in it. The landscape does not express the mood; it rather has it. The mood surrounds, fills, and penetrates it, much like the light in which it shines, or the scent that emanates from it. It belongs to its overall impression and can only be released from it as a special part through abstraction. (Schmitz 1969, 368-370).[2]

At this point, the distinction between Klages' expressive-(physiognomic-) phenomenological approach and Schmitz' felt-body-phenomenological approach becomes clear. For the expressive phenomenological perspective, the eye – in its unreduced, fully sensuous and immediate capacity for seeing and observing – serves as the distinguished and preferred "organ" for disclosing the world. In contrast, for the felt-body-phenomenological approach, it is the unreduced, inherent bodily sense. Therefore, when Klages states: *"There are as few sensation experiences free of feelings as there are feelings entirely free of images,"* (Klages 1936, 95; cf. Kozljanič 2004b, 262-287 and Kozljanič 2024a) he is correct from the standpoint of expression phenomenology. Likewise, when Schmitz – from the corporeal phenomenological perspective – asserts that an "objective feeling" is a purely spatial, atmospheric, object-related phenomenon, and that this phenomenon does not express any kind of animated "being" or "image," but is merely an emotion "condensing" around an object: then he is equally right. For corporeal sensing, through its very act of sensing, genuinely captures only the feeling itself – not the physiognomic, face-like expression.

According to Schmitz, an "objective feeling" presents itself "as a floating, elusive echo or 'whiff' around certain objects." (Schmitz 1969, 376) When this "whiff" condenses into a meaningful atmosphere, so that it appears almost like a "being" (or also a "soul" or "image" in Klages's

---

2 Schmitz took the Baensch quote from: Baensch, O. (1924). Kunst und Gefühl. In Logos, Vol.12, 14 and 2.

sense), the view arises that it is actually something like an autonomous, animated "being." However, such talk of "being" is misleading. Schmitz therefore recommends: "To advocate the conception of beings as objective, incompletely intentional feelings." This view becomes clearer when considering Schmitz introducing the distinction between "emotional expression" and "experiential expression." For now, it suffices to record his provisional summary: it is "more precise to specify the talk of beings as a subclass of objective emotions." (Schmitz 1969, 376f.). So far, so good; however, when Schmitz attempts to discredit the expressive phenomenological and physiognomic approach by accusing it of failing to provide a "satisfactory, terminological version of the intended," claiming it as "vague" and "mysterious" (Schmitz 1969, 375-377), and that it is also subject to a "phenomenologically insufficiently developed language" (Schmitz 1969, 369), then this is not correct. Schmitz clearly oversteps his bounds here and overlooks the fact that his corporeal phenomenological approach is only one of several permissible phenomenological approaches, and by no means *the* definitive phenomenological method. He neglects the fact that – albeit from a somewhat different perspective – the phenomenon in question can also be clearly described and clarified through expressive phenomenology (cf. Kozljanič 2004b, 262-300 and Kozljanič 2024a).

Elsewhere, in his study "On embodied communication", Schmitz dealt more extensively with expressive phenomenological theory. There, too, he attempted to translate expressive phenomenological achievements into corporeal phenomenological terms. And there, too, the translation was not entirely successful, because, as stated, one approach cannot be completely resolved into the other. In the aforementioned study, Schmitz introduces the distinction between "experiential expression" and "emotional expression". By emotional expression, he means something that is expressed in an "objective feeling": his peculiar atmospheric quality. By "experiential expression", on the other hand, he designates something that can only occur in humans and animals and that testifies to an "affective involvement", to some kind of inner experience on the part of these living beings.

Emotional expression "appears in many places in the world where there can be no talk of the expression of experiencing. The howling wind at night can lament as well as a human or a violin, which can in addition rejoice and sob, but wind and violin experience nothing; nevertheless, the emotional expression in their sound is just as genuine as in that of the voice." (Schmitz 1992, 184; cf. Schmitz 1978, 75-109). What Schmitz described as quasi-essential, incompletely intentional, objective emotions could be roughly identified as emotional expression. All "fully intentional feelings" would then correspond to the experiential expression.

Therefore, actual bodily communication, a bodily dialogue, can only occur in relation to carriers of experiential expression, which means in relation to humans and animals. Schmitz illustrates this with an example of a rider who, while galloping, "fuses" in a way with the horse, as Schmitz puts it "embodies" herself with the horse. This "fusion" resp. "embodiment" or "encorporation", is not absolute but dialogic, of course. The rider does not wilfully succumb to the horse, but rather provides the initial and propulsive impulse to gallop. The horse, for its part, does not react mechanically, like a car due to the depressed accelerator, but as a living being in its own right. It "not only carries out its task, but also reflects the impulse to do so onto the rider, who rediscovers her own urge and will in its movements and thus bound to it even more. In an interaction of initiative and resonance, rider and horse pass on the shared impulse to move to one another." This dialogical, corporeal communication between people is even more evident, for example, in erotically encouraging interactions or in a threatening, combative confrontation.

> In contrast to the contemplation of an [emotionally] expressive natural scene, e.g. a dark cloud formation clustered over mountains, here a dramatically-dialogical relationship arises between the participants: Each sounds out the partner, as it were, and measures them by expressively sending impulses to them, which the other returns strengthened, weakened or modified, so that each becomes aware of what they are dealing with through the modification of their own corporeal state. (Schmitz 1992, 195f.)

As I attempted to show in my monograph "Der Geist eines Ortes" ("The Spirit of a Place") this form of dramatically-dialogical expressive communication also exists among mythical peoples in relation to beings that are neither human nor animal (cf. Kozljanič 2004b, 279-295). The entire spectrum of "imitation magic" (i.e., the so called "image and sympathy magic") testifies that this type of communication is also possible and meaningful with plants, stones, clouds, mountains, rivers, etc.; it testifies that even a supposedly "inanimate thing" sends back to the mythic person the impulse sent out ritually and imitatively, "strengthened, weakened or modified, so that each [of the people participating in the ritual] becomes aware of what they are dealing with through the modification of their own corporeal state." Accordingly, the distinction between emotional expression and experiential expression would be invalid or at least questionable. One would have to make a different distinction – as Klages did – namely, that between direct, involuntary expression of the true nature of a being or "quasi-thing"[3] (which can appear in humans, animals, plants, stones, places, landscapes) and represented, played, "pretended" arbitrary expression. Only humans are capable of arbitrary expression, for only humans can "disguise" themselves in the full sense, i.e. consciously.

Schmitz, however, insists on the distinction between emotional expression and experiential expression and comes into obvious contradictions: On the one hand, he formulates that with the reciprocal dialogical, corporeal communication he "found the phenomenological criterion for the above mentioned distinction between emotional expression and experiential expression"; this insofar as such a dialogue is only possible between partners capable of experiential expression. On the other hand, he must add in the following sentence: "Phenomenologically speaking, the difference, of course, does not depend on whether the partner in whom the experiential expression is perceived is truly an experiencing being." This means, said dialogue would also be possible between a human and a plant or a prebiotic "being". However, this abolishes or at least greatly relativises, the "phenomenological criterion for the distinction made above between emotional expression and experiential expression." This is probably also the reason why Schmitz remains comparatively vague regarding the distinction between experiential expression and emotional expression: "With the mere emotional expression, which is not an experiential expression the relationships are similar, even if the reciprocity of embodiment [resp. bodily communication] is then absent or completely recedes into the background." (Schmitz 1992, 197, 199).

4.3 *Local atmospheres as locally-spatial anchored and object-bound emotions*

Climatic, climate-like atmospheres and landscape atmospheres are, like all other atmospheres, too, in Schmitz' words "Spatially extended, each time borderlessly outpoured, corporeally sensed emotions". What turns out to be special about landscape atmospheres is that, they are object related and not merely spatially extended outpoured. Their spatial

3 The term „quasi-thing" (or „half-thing") was coined by Schmitz (cf. 1995, 216-222) and further developed by Tonino Griffero (2017).

extension has, so to speak, contracted and scenically condensed and aligned, i.e. structured. In other words: A directionally spatial layer built up over its spatially extended layer. In case of a further compaction and concentration the atmosphere aggregates in one place or even in an individual "object" present at the location, then we could speak of "local atmospheres". With these atmospheres a third spatial layer comes up: Schmitz calls it the "locally-spatial" layer. "On the basis of the corporeal directional space, the "local space" arises when relative [extracorporeal and more precisely definable] locations are introduced via the positional and distance relationships between the targets of corporeal directions." (Schmitz 1995, 284).

As far as I know, Schmitz does not use the term "local" (or locally-spatial) atmosphere. However, he knows and names the phenomenon. After presenting the case of a human affected by a winter landscape atmosphere, he says:

> While the objective emotion in this case may have been evenly and widely spread [over the nocturnal winter landscape], it occasionally contracts around certain objects as well, which it surrounds and distinguishes like a nimbus. Osvald Sirén experiences such a feeling as a numinous power in a Chinese Buddha statue from the Lung Men caves: 'Anyone who approaches this figure will realize that it has a religious significance, without knowing anything about its motive. It matters little whether we call it a prophet or a god because it is permeated by a spiritual will, which communicates itself to the beholder. The religious element of such a figure is immanent: it is a 'presence' or an atmosphere rather than a formulated idea.' In this case the powerful, immediate presence of the numinous dispels the suspicion of a mentally transmitted transference [read: projection] of the objective feeling onto the object and reveals its atmospheric nimbus as an intrusive date to everyone, provided the source can be trusted. (Schmitz 1969, 370f.)

Local atmospheres are therefore denser and more centred than landscape atmospheres. Their expansiveness has contracted and is, as it were, centered on one place. Just as a climatic atmosphere is characterised by an undifferentiated, purely expansive diffusion, and a landscape atmosphere by an already structured, directional and not merely expansive diffusion, the local atmosphere is characterised by a further concentration of spatial effusion. The local atmosphere can thus be anchored in an 'object'. It concentrates around a very atmospherically dense place with a comparatively clear locally spatial structure. The density and centeredness of the local atmosphere can culminate in an "auratic object". Such an "auratic object" on which an atmosphere condenses and anchors itself, can be, as in the above example, a Buddha statue in a cave, but just as well an old tree, a strange rock, a secret spring, a waterfall or a striking rapid etc.

To understand this process of condensation and object centeredness even more deeply, a distinction may be helpful, which Schmitz adopted in a slight modification from the gestalt and expression psychologist Wolfgang Metzger: The distinction between the "area of condensation" and the "anchoring point" of a "gestalt". What does this mean?

> The *area of condensation* of a gestalt [or in Klages' terminology, of an "expressive image"] is the place where its character visually gathers, e.g. regarding the facial expression it usually is the area around the eyes or the outline of a leaf or a leafy branch. Metzger on the other hand, refers to the *anchoring point* as 'that point from which the whole appears to be constructed, and which therefore represents the whole in its environment'; it is to be found in many gestalts, e.g. in a drop, tree and shell, 'at the end opposite the area of condensation'. (Schmitz 1969, 317; cf. Metzger 1941, 171, 174, 176).

Schmitz tried to make this gestalt psychological finding fruitful for his New Phenomenology. Using a fairy tale by the Grimm Brothers (tale of the Robber Bridegroom), he tries to demonstrate the suitability of the terms "area of condensation" and "anchoring point" in relation to his concept of atmosphere. A young woman is married without knowing her groom, where he comes from and where he lives. From the very beginning, she has an uneasy feeling about it, an uneasy feeling incidental to her unknown groom. This feeling intensifies even more when she enters into the groom's house lying seemingly deserted in the forest. It becomes completely anchored and centered when a voice calls her out to her in this house: "Turn back, turn back, you young bride, you are in a murderer's house." Schmitz observes:

> The bride's horror in the fairy tale already has a condensed area (groom, house), but not an anchoring point yet; thus it is object-related in one sense, but objectless in another. Only at the moment when it gains an anchoring point, namely the danger that she will be murdered (if she does not get into safety), does the pure diffuse excitement become centered fear. (Schmitz 1995, 301).

If I understand Schmitz correctly, the situation could be interpreted to mean that a landscape atmosphere can be more precisely understood as an atmospherically condensed area, and a local atmosphere resp. an "auratic local object" – insofar as something atmospheric condenses and concentrates at specific points – more accurately comprehended as an atmospheric anchoring point. Thus, the following connection between climate-like, landscape and local atmospheres would be conceivable: a climatically spatially extended outpoured atmosphere could condense into a landscape spatially-directed atmosphere and this, in turn, could anchor itself at certain points at places resp. locally-spatial atmospherically-auratic "objects".

I am not sure whether Schmitz would endorse my interpretations. However, I am sure that this would allow us to clarify and deepen much of Heidegger's and Norberg-Schulz' genius loci concept.

Here is an explanatory side note. As I was able to show in "Der Geist eines Ortes" (Kozljanič 2004b, 315-320, 342-344) it makes sense to distinguish between three aesthetic spatial levels. Firstly, the *place*: a rather small-scale level, a single, clearly defined, aesthetically appealing place with its small ensemble of things and living beings. The ideal type here would be the idyllic place, the 'locus amoenus' with the classic ensemble of rock, spring and tree. Secondly, the *landscape*: a rather medium-sized, panoramic, easily comprehensible area, staggered into foreground, middle ground and background. The ideal type here would be the picturesque landscape of 19th-century painting in its two main aspects of the 'beautiful' and 'sublime' landscape. Thirdly, the *earthscape*: a large-scale (no longer visible from the ground) geographical-climatic unit formed by striking deserts, high mountain ranges, vegetation belts, sea areas, large plains, coastal ranges or even polar zones. The ideal type here would be 'photogenic geozones' as they can be admired in illustrated books with titles such as "The Earth from Above" or "The Earth from Space".[4] However, geozones are not only accessible to us technically from an aeroplane or even satellite perspective, but also without such aids, sensually and aesthetically from the ground: namely through their climate and their sky and cloud conditions; a desert sky is experienced atmospherically differently than a high mountain sky or a tropical sky.

In "Der Geist eines Ortes" (Kozljanič 2004b, 315-320) you will find my reference to Christian Norberg-Schulz's three major earthscape types (Norberg-Schulz 1982, 42-47): northern romantic, Mediterranean classical, desert cosmic; and the "climate zones" of Watsuji Tetsuro (1997, 21-104): monsoon, desert and meadow climate. In general, I believe that the lists provided by Norberg-Schulz and Watsuji are far from sufficient. There are many more relevant earthscapes and/or geozones to be distinguished here.

4 Satellite photography now also offers the possibility of photographing geozones from space, thus making them accessible for aesthetic contemplation; see, for example: Arthus-Bertrand, Y. (2014). *Die Erde vom All.* München: Knesebeck.

One can describe, with this distinction in mind, earthscapes *in the sense of the most general climate-like atmospheres as purely spatially extended perceptible phenomena; can further describe the gathering "potency" of an area (or, with Heidegger, Gegnet and landscape; see Chapter 2 as a directional spatial process of condensation and a perceptible object-related phenomenon; last but not least can describe the gathering "potency" of a place and building as a locally-spatial process of anchoring and as a perceptible object-bound phenomenon.*

As a result a "Mediterranean classical" *earthscape* would be condensed in a Greek (divine-)landscape – such as the Apollonian landscape around Delphi or the Athenian landscape around Athens. The (divine-)landscape, for its part, would then be anchored in daimonic places – in Delphi, for example, at the spring of Castalia or the Temple of Apollo, in Athens, for example, in the sacred olive tree or at the Parthenon on the Acropolis. Or to summarize the whole thing using a more secular example: a "northern romantic" *earthscape* could condense itself to a Norwegian fjord landscape anchoring itself at a mighty waterfall pouring into the fjord; it could condense in a Middle Rhine landscape anchoring itself at a rock with a castle ruin; or it could condense itself in a German low mountain range landscape in order to anchor itself in a mighty fir or a rocky promontory, too.

My studies have often shown that these considerations about the connection between earthscape, landscape and place feature an undeniable proximity to phenomena. But I don't want to withhold another example. It comes from Adalbert Stifter (1805-1968), *the* poet of the Bohemian Forest. He often described Lake Plöckenstein, located at the foot of the highest peak, the Plöckenstein in the southern Bohemian Forest, as the heart and eye of this landscape – quasi its anchoring point.

One of these descriptions can be found at the beginning of the narration „Der Hochwald" (The High Forest). Stifter describes the wooded mountain range, which rises to the height of the Plöckenstein peak as sylvan "wall"; this "wall" pursuant to Stifter

> is our most splendid focus. The true point, however, is a lake, which it supports approximately at the second third of its height. [...] A feeling of the deepest loneliness overcame me almost invincibly, however often and gladly I climbed up to the fabulous lake. [...] One and the same thought often emerged from me, when I sat at these shores: – as if it were an uncanny natural eye looking at me here – deep black – towered by the front and the brow of the rocks, fringed by the eyelashes of dark firs – the water within motionless like a petrified tear.

[Pic. 29: Plöckenstein Lake in the Bohemian Forest]

From this example as well, it is clear without too much effort to interpret: earthscapes condense in landscapes. Landscapes are anchored in places. If this is the case, however, it has consequences for the architectural and creative approach to places. For where places resp. landscape anchor points are not treated with care, there is a danger that the auratic "nature" of a place will dilute atmospherically or vanish. However, if landscapes gradually lose their anchor points, then the landscape itself is in danger of losing its specific quality as a condensing atmosphere and generating into a comparatively expressionless and characterless "region". And there are already too many "regions" of this sort today. The primary goal of a genius-loci-conscious approach to places should therefore be to limit excessive planning relations (overplanning) to already de-densified landscapes, i.e.: "regions", to proceed with extreme restraint with relations of overplanning, and to favour relations of structuring in condensed areas and relations of integration at anchor points.

All these deliberations and conclusions show that the Schmitzean concept of spatially extended atmospheres allows us to see and describe the phenomenon of the genius loci, in many respects, more comprehensively and differentiated than previous concepts. However, there is the risk that: the danger of a "Platonising" hierarchy of the various genius loci levels. For we should be clear about this: Schmitz conceives the condensation and anchoring process as *ultimately* hierarchical. Initially, as the first, original, paradigmatic, and general, there is the climate-like, pure spatially extended atmosphere. In contrast to it, landscape condensed and locally anchored atmospheres (at least in tendency) already represent something secondary, derivative and "inferior". They only exist "by the grace" of a more general and comprehensive atmosphere. But this hierarchy, and especially the evaluation associated with it, is more than questionable. We could just as well advocate a contrary hierarchy and evaluation and ask: Is a landscape without auratic and strongly atmospheric places a landscape at all, and not rather merely an expressionless and characterless 'region' and an insignificant planning area? Does not landscape only become possible with such places and through places? So are places not rather the original and landscapes already a derivative? And: Is not earthscape only possible through such place-bearing landscapes? Or also: Is earthscape, which qua earthscape represents an incomprehensibly large-scale phenomenon, still experienceable without place-bearing landscapes, without landscapes in which we first gain an inkling of what earthscape can be? In addition to the two opposing hierarchies, a third option seems possible: a non hierarchical approach that recognises, explores and describes each of the three genius-loci-levels in their phenomenal existence. I advocate this option.

4.4 *Local daimons as personified and specified local atmospheres*

According to Schmitz, as before mentioned, there are not only supra-personal, mundane, but also supra-personal "divine atmospheres". The atmosphere of "Christian love" or of the "Holy Spirit", for example, the ecstatic Dionysian atmosphere, the thoughtful and reserved Athenean atmosphere: All these are supra-personal "divine atmospheres" which Schmitz analyses and attempts to define using his felt-body-phenomenology (cf. Schmitz 1977, 12-19). What Schmitz calls "local, divine atmospheres" also falls into the category of "divine atmospheres". A numinous place experienced purely in terms of atmosphere in the ancient sense: This is Schmitz' paradigm of such a local divine atmosphere. It is not for nothing that he uses the same Seneca quotation at a central point that I also used in Chapter 2 (Seneca. *Ad Lucilium Epistulae Morales,* 4, 41, 3; cf. Schmitz

1977, 130). This quotation is more suitable than almost any other for illustrating the atmospherically numinous ancient mode of experiencing place.

More interesting in this context, however, is a quotation from Pomponius Mela's "De Chorographica" (1, 13, 72-75), cited and (I assume) also translated by Schmitz:

> Up there is the Corycian cave, with a unique atmosphere and magnificent almost beyond any description. Directly at the highest peak of the mountain bordering the shore and being quite steep with a slope of 10 stades (approx. 1850 metres), it gapes and opens it up with a huge cleft. Deeply sunk, it grows then ever more lush with groves covering the slopes all around, and the whole is enclosed on the sides by a ring of forests: so wonderful and beautiful that it momentarily unsettles the minds of those who approach, where even lingering in contemplation will not satiate them [...] Inside the space is so terrifying that no one dares to enter it, and therefore it remains unknown. The entire cave, however, is solemnly sublime and truly holy, worthy of being inhabited by gods, and this is also believed. Everything is venerable and reveals itself as it were, with a kind of numen. (Schmitz 1977, 132).

[Pic. 30: The famous Corycian Cave. Next to Mount Parnassus, above Delphi.]

[Pic. 31: The interior of the cave is full of large and small "wonders".]

The quotation is interesting because it shows how Schmitz attempts to limit the full archaic-mythical phenomenon of the genius loci – which consists not only of the numinous experience in the form of atmospheres, but just as much, if not more, of the expressive image and essence in which the local daimon appears (see Chapter 2) – to the felt-body-phenomenologically knowable atmospheric half. For in the Latin text of Pomponius Mela, it does not read "Up there is the Corycian cave with a unique *atmosphere*" but: "Up there is the Corycian cave, with a unique *ingenium*". In this context the Latin word "ingenium" rather suggests a local genius and "nascent spirit" that acts autonomously and is certainly not purely atmospherically present. Schmitz obscures this extremely important aspect. The reason for it is to be found in the specific hierarchies and evaluations of his phenomenological approach. In case of doubt, his body-felt-phenomenological approach considers the mood-related, emotional, atmospheric (because this is precisely what is initially and primarily present in felt body field) to be the original, while the figurative and pictorial are "merely" derived. This methodological preference for the purely mood-related was already evident in point 4.1, where Schmitz established the corporeal sensing of climate-like atmospheres as *the* paradigm for experiencing of all supra-personal feelings, then in order to derive, as presented in 4.2 and 4.3, from this "archetypal phenomenon" the "more concrete", "object-related", and "object-bound" landscape and local atmospheres as "condensation and anchoring" processes.

This preference and hierarchisation is now also evident in Schmitz' concept of "divine atmospheres" And most clearly so in his corporeally phenomenological working hypothesis hereof, stating "that every god is an emotion" (Schmitz 1969, 128). Every god: i.e. also every daimon, i.e. also every local daimon. Against the background of this working hypothesis, it is not surprising that Schmitz comes to the conclusion

> that the local divine atmospheres belong to the immense realm of supra-personal and objective emotions, which are partly placeless, as it were, 'in the air', or more precisely in the extended space [...] like the weather, and partly condensed in specific places and around specific objects, often only in the manner of floating echoes like the [Quasi-]beings (Schmitz 1977, 133f.).

Thus, not only the landscape and local *mundane* atmospheres, but also the local divine atmospheres and with them all other divine atmospheres, are "nothing but" derived resp. condensed and anchored climate-like and spatially extended feelings. Against this background Schmitz can state quite basically:

> Hence the authority of the divine is, at its root, always the authority of emotions [...] ; it can also appear as the authority of persons (gods) insofar as divine atmospheres, which are emotions, quasi condense in these persons [...], but their divine authority then goes back to the empowerment by these atmospheres (Schmitz 1995, 439.)

If this is the case, however, then the question arises, of course, how it can come about that gods and daimons also appear in a pictorial manner or resembling a human form, hence person-like condensed and person-like anchored? According to Schmitz, through acts of personifying "interpretation" and "reification". What does he mean by that? The following: Since the divine makes itself originally noticeable only as a climate-like, spatially extended atmosphere, it is only vaguely tangible and nameable for the people to whom it happens atmospherically. This vagueness, however, is not sufficient for people: They want to communicate about the divine, want to understand and clearly name it. In short: They want to concretise it, visualise it, personify it, and anthropomorphise it. Furthermore, in some divine-atmospheric experience,

> a pressure towards concretisation is almost persistently noticeable, without succeeding in satisfying it; the tormenting question remains: What was it after all? It happens, however, that concretisation can be achieved at the price of personification, and the cases are of crucial significance for the transition from the divine [as pure divine atmosphere 'being in the air'] to the [concrete, because personified] god. (Schmitz 1977, 149)

Through this process of concretising and personifying "interpretation", not only clearly defined gods are formed, but also local daimons. Here, too, Schmitz establishes a human tendency and urge to the reification of divine atmospheres:

> Similarly the atmosphere of twilight fear without figures like the Lorelei and her Indian sister Aranyani, the enchanting and seductively captivating power of the quiet sultriness, would not be concrete without the sirens as personified incarnations and interpretations of this highly emotionally powerful atmosphere [...]. Only through such personifying incarnations and through them the talk of such powers of nature gains the force of referring to the concrete, whereby all that is pale [...] and the uncertainty of the imprecise declines from them, and an unmistakeable face of reality, which draws its suggestive force from the nimbus of a powerful atmosphere [...] emerges. (Schmitz 1977, 150)

No matter how hard Schmitz tries to exhaustively elucidate from a body-felt-phenomenological standpoint the phenomenon of local daimons: He only ever succeeds in addressing the atmospheric side; the other crucial aspect – that local daimons are not "interpreted" and hence derived phenomena for mythical people, but rather, entirely in the Klagesian sense, image and essence-like appearances – remains phenomenologically ungraspable, despite all attempts to make them plausible as "concretion" and "personification". Schmitz is aware of this and helps himself with an embarrassing argument that he puts forward prior to all his explanatory attempts.

> How does such a personification arise again and again, at all times and among all peoples, when atmospheres are hardly persons in themselves? [...] This brings us to the question of truth, which is decisive for religion in the usual sense and to divine persons in the related sense. It cannot be conclusively answered by phenomenology alone; for belief in divine persons, understood as a metaphysical conviction, can as little as be definitely confirmed or refuted purely phenomenologically than beliefs in fairies, witches, dwarves, Goethe's mothers or Kant's homo noumenon and other conceivable concoctions. (Schmitz 1977, 146)

Of course, what Schmitz calls the “question of truth” of religion can never be definitely answered, confirmed or refuted by phenomenology (as by any other science and philosophy). And of course, we can argue forever about “metaphysical convictions” and all kinds of “concoctions”. But this is precisely what phenomenology is not about. And Schmitz knows it, too. Phenomenology is concerned with the recognition, exploration, and description of the full, multi-layered, unreduced, undisguised phenomenal reality. And on this level of experience, gods and daimons reveal themselves as genuine and more-than-merely atmospheric phenomena. It would have been more modest and appropriate if Schmitz, instead of a sweeping and ultimately polemically embarrassing argument, had resorted to a methodologically restrictive explanation. For example, that through his felt-body-phenomenological method, the atmospheric dimension of the daimonic and numinous can be demonstrated, differentiated, and described with particular clarity, while other dimension of the same phenomena can only be revealed vaguely, deficiently or not at all. His grandiose and pioneering new phenomenological insight into the “world of atmospheres”, his perspicacious, sensitive, and conceptually highly differentiated description of atmospheric phenomena that are spatially extended, spatially directionally condensed and spatially locally anchored, would have lost none of their phenomenological content; quite the opposite.

Moreover: Only against the background of such a methodologically restrictive explanation could we have seen what Schmitz actually illuminates with his theory of concretion and personification: namely, the process of aesthetically-metaphorical image intuition described under point 7.1.4 in my monograph “Der Geist eines Ortes” (Kozljanič 2004b, 295-300). What is not true for the genuine mythical vision of daimons and gods – that these were processes of concretion and personification – is certainly true for aesthetically-metaphorical image intuition. What I said under the aforementioned point 7.1.4 could be further specified as follows: In aesthetically-metaphorical image intuition, reality appears atmospheric, deep, numinous, and auratic but not as a daimon. And because it is not daimons that appear, and because the appearance, as a mere atmospheric one, would remain too vague and intangible, the artist himself helps out and forms or creates an “allegorical” image, a metaphor in which the phenomenon, which in the mythical world was still allowed to “concrete” itself, can now only show itself “concretely” in the metaphorical mode of “as if “. The concrete daimonic image, which the modern artist, qua artist, was unable to evoke from reality is replaced by an aesthetically interpreted self-creation – a concretising and/or personifying metaphor. The true mythical vision gives way to artistic metaphor, the visionary seen to the poetically imagined, and the "reality of images" is replaced by a "reality of metaphors" as the condensation and concretion of powerful spatially outpoured atmospheres.

4.5 *The genius loci as the archetype of a place?*

The only phenomenologist who has done justice to the daimonic, personal dimension of the 'genius loci' that Schmitz (and, by the same token, Rudolf Otto) overlooked is Ludwig Klages (1872-1956). For Klages succeeded in throwing light on such phenomena as the 'genius loci' not only in their atmospheric and numinous, but also in their daimonic and essential aspects. Without being able to go here into detail about his doctrine – which, in some respects, is close to Jung's notion of archetypes, if in a less Platonic and more Heraclitean way – of the 'reality of the (primordial) images' ("*Wirklichkeit der* [*Ur*]*Bilder*"),[5] I should like to refer to it using a rather long, but highly suggestive, passage, in which Klages explains how "the *dimension of the daimonic* is essentially identical with the temporal-spatial uniqueness of an image":

> For us [i.e., for modern, "intellect-oriented" humans] there exists [three-dimensional] space and bits of space, [linear] time and bits of time; for the life-bound [i.e., biocentric] spirit [of mythically-minded humans] there are as many spaces and times as there are images to be spatially and temporally experienced: there is a nocturnal space and a diurnal space, a domestic space and a cosmic space, a forest space and a temple space, an eastern space, a southern space, western space, northern space, but in certain circumstances there are as many house spaces as there are different houses, and even as many different spaces in one and the same house as there are moments of dynamic internalization of spatial appearances that constitute the house [*Augenblicke wirksamer Innerung der hausgestaltigen Raumerscheinung*]. The earliest form and the root of knowledge of the reality of images [*die Wirklichkeit der Bilder*] is polydaimonism, and the multiplicity it announces multiplies every content of meaning [*die mit ihm sich bekundende Vielfachheit vervielfacht jeden Bedeutungsinhalt*]. (Klages 1969, 1263).

5 For further discussion, see the chapter "Das ausdrucksphänomenologische Konzept der ‚Bilder' und ‚Elementarseelen' eines Ortes von Ludwig Klages" (Kozljanič 2004b, 262-300).

Klages goes on to explain what, in the context of his doctrine of the 'reality of the images', he means by talking about daimons:

> The true daimon is the daimon of a place, a landscape, an element, which changes with the way their own appearance changes; which receives the offerings of those who acknowledge it in temporal rhythms, whose daily, monthly, or annual patterns are based on the forms in which it is revealed; and which is at work in those events or tasks that the [mythically-minded] spirit was compelled to accept as powerfully at work through their images, of which pathic [and visionary] perception is aware. (Klages 1969, 1264).

If Klages is right in his assessment, and there is much to suggest he is, then one could discern in the 'nascent spirit' of a place as something archaic and originary one last aspect, and perhaps its most profound: namely, that at its heart lie powerful and evolving (primordial) images. The genius loci would thus not only be a tutelary and protective spirit and a 'spirit of becoming', not only a 'spirit of fate' and a 'spirit of development' (*Bildegeist*), but precisely also a 'spirit of images' (*Bilder-Geist*). Or, following C. G. Jung's formulation, originary sites of a 'genius loci' would be inhabited by archetypes, such sites would possess an innate archetypal potency. And the numinous atmosphere of a site would be, in the final analysis, the aura or the charisma, the herald or the harbinger of a local archetype, an *Ortsarchetypus*, that is surging into appearance.

For our modern, rationalistic world such thoughts are, of course, nothing less than heresy. But it is not only Klages who entertains these ideas: C. G. Jung (1875-1961), too, is attracted to them, most often with reference to the comparative religious concept of "mana", which is "not confined to Melanesia, but can also be found in Indonesia and on the east coast of Africa", and "still echoes in the Latin *numen* and, more faintly, in *genius* (e.g., *genius loci*)" Jung (1969b, 233). If, as Jung says, the psyche is "a reflection of the world and man"; if, precisely for this reason, it is "the only phenomenon that is given to us immediately and, therefore, is the *sine qua non* of all experience" (Jung 1969a, 139); and if, on the level of this world-related psychic experience there is much to indicate that the "mana"-concept is not quite so unreal as today we moderns would like to believe, then "we hesitate to accept this and begin to look around for a comfortable theory of psychic projection" (Jung 1970, 60).

For "the question", Jung continues, is this: "Does the psychic in general — the soul or spirit or the unconscious — originate in *us*, or is the psyche, in the early stages of conscious evolution, actually outside us in the form of arbitrary powers with intentions of their own, and does it gradually take its place within us in the course of psychic development?" (Jung 1970, 60).

However one chooses to answer this question, whether one sees in the site of a genius loci the presence of an archaic and originary (mana) spirit or whether one sees it simply as a surface onto which something is projected,[6] one thing remains certain: it is not only in relation to the path of life in the biography of an individual that a certain originary magic becomes noticeable, but in relation to important locations in the social space of life as it is lived. Indeed, one could take those lines from Hermann Hesse's poem "Steps" ("Stufen") – "For in *every beginning* lies a special magic / That nurtures life and bestows protection"[7] – and, in the light of the concept of the genius loci, modify them as follows: 'For *certain places* have a special magic, / That protects it and helps us to live.' For such archaic and originary places really do help us: they renew our connection with our natural, social, and personal, life-historical roots, and so make possible a firm foundation for a future, in which continuity and transformation are not seen as opposites, but as reciprocally interrelated and as establishing the place we call our home.

---

6 On the depth-psychological theory of projection, see Franz, M.-L. v. (1978). *Spiegelungen der Seele: Projektion und innere Sammlung in der Psychologie C. G. Jungs*. Stuttgart: Kreuz, 11; see also Jung (1969c, 25f.). This theory has been applied to the problems of landscape and urban development; see Abt, T. (1983). *Fortschritt ohne Seelenverlust: Versuch einer ganzheitlichen Schau gesellschaftlicher Probleme am Beispiel des Wandels im ländlichen Raum.* Bern: Hallwag, 149. A similar theme is discussed (although on the basis of Freudian psychoanalysis) in Lorenzer, A. (1968). *Städtebau: Funktionalismus und Sozialmontage? Zur sozialpsychologischen Funktion der Architektur*. In H. Berndt, A. Lorenzer, K. Horn (eds). *Architektur als Ideologie*. Frankfurt a. M.: Suhrkamp, 51-104 (p. 83); see also C. Majunke (1999). *Der* Genius loci: *Geist des Ortes oder verorteter Geist: Landschaftsplanung zwischen dem Wunsch nach Ganzheit und moderner Subjektivität*, Berlin: Eisel, 1999, 89; and Kozljanič 2004b, 370-83.

7 In German: „Und jedem Anfang wohnt ein Zauber inne, / Der uns beschützt und der uns hilft, zu leben." H. Hesse, H. (1987). *Gesammelte Werke in zwölf Bänden*. Vol. 9. Frankfurt am Main: Suhrkamp, 483. The poem appears in Hesse's novel *The Glass Bead Game* (*Das Glasperlenspiel*).

5.
Walter Benjamin's Concept of Aura and the Aura of a Place – By Reference to Judengasse in Frankfurt

In terms of reception history, the concept of aura is closely linked to its thematisation in Walter Benjamin's essay "The Work of Art in the Age of Mechanical Reproduction" (Benjamin 1969 & 1980). Looking back from Benjamin's essay, we arrive at Ludwig Klages and his concept of "primal images",[1] which in a certain sense, was the force behind Benjamin's concept of aura.[2] Looking forward from Benjamin's essay we arrive at further literature,[3] from which I will select two examples: a book by Gernot Böhme (1995) and an essay by his student Michael Hauskeller (1999).

Hereafter, I will first present Benjamin's concept of aura. Then I will discuss the explications and modifications this concept underwent by Gernot Böhme and Michael Hauskeller. After these conceptual definitions, I will focus my attention to the life-spatial phenomena of aura – more precisely the "aura of a place": To this end I will turn to the experiencing and the experience some people had with the ruins of the Judengasse in Frankfurt. Finally, I will attempt to grasp the phenomenon of the aura of a place more sharply and nuanced than before and, at the same time, to show the symbolic function of such places in its true light.

---

1 "Primal images" are 'deep-reality-images' ("Urbilder"): cf. Klages (1929, 218-221); Klages (2023, chapter 4f.); Kozljanič (2004c, 178-182); Bishop (2018, 73-88).

2 On Klages' influence on Benjamin, see: Pauen, M. (1994). *Dithyrambiker des Untergangs*. Berlin: Akademie, 187-189; also insightful on this topic: Morita, D. (2006). *Von der Phantasmagorie zum dialektischen Bild. Walter Benjamin und das Passagen-Werk*. In H. Paetzold (ed.). *Integrale Stadtkultur*. Weimar: Verlag der Bauhaus-Universität, 30-46.

3 See, for example: Stoessel, M. (1983). *Aura. Das vergessene Menschliche. Zu Sprache und Erfahrung bei Walter Benjamin*, München: Hanser; or: Recki, B. (1988). *Aura und Autonomie. Zur Subjektivität der Kunst bei Walter Benjamin und Theodor W. Adorno*. Würzburg: Königshausen & Neumann.

5.1 *Benjamin's concept of aura*

Benjamin defines the aura of natural objects as a "*unique phenomenon of a distance, however close it may be*" (Benjamin 1969, 5). To clarify his concept of aura Benjamin refers to a contemplative experience. "While resting on a summer afternoon and following a mountain range on the horizon or a twig that casts its shadow on the resting person – this is to breathe the aura of these mountains or this twig." (Benjamin 1980, 479). This short definition and the subsequent example alone show what Benjamin imagines by aura: namely nothing religious or even "esoteric" in the narrower sense, no glaringly shining nimbus, halo or gloriole surrounding the respective auratic objects, but something else: an original perceived in a resting and contemplative way,[4] in an aesthetically watched distance[5] resp. remoteness that is atmospherically charged and expressive.

With the term "aesthetic devotion" coined by Hermann Schmitz, what is ultimately described here quite vaguely as "aesthetic distance" resp. "remoteness" is now defined more clearly.[6] A contemplative view and

4 Benjamin also refers to this as "free-floating contemplation" and "contemplative immersion".

5 Benjamin's concept of "distance" is clearly influenced by Klages' concept of "eros of distance" (a distance in which the 'images' of apparent reality light up); cf. Klages (1974 & 2023, chapter 4f.); and Benjamin (1961, 327f.) – Benjamin knew and appreciated Klages' writing. In a letter to Klages dated 28 February 1923, he wrote: "I take the liberty of expressing to you, on the occasion of these lines, the joy and confirmation of my own thoughts that I have gratefully derived from your writing on Cosmogonic Eros." Quoted from: H. E. Schröder (ed.) (1972). Ludwig Klages 1872-1956. Katalog zur Centenar-Ausstellung. Bonn: Bouvier, 165.

6 Hermann Schmitz attempted to distinguish „aesthetic devotion" from both "religious devotion" and "aesthetic enjoyment": "Aesthetic and religious devotion are closely related. [...] Purely religious devotion [...] is immediate emotion evoked by the presence of the divine; it lacks the aesthetic distance, the mediating power of aesthetic devotion, which grants and imposes on the personally emancipated human being [...] indirect emotion from a powerful atmosphere [such as a scenic mood] without exposing him to the naivety of immediate and unreserved surrender to this atmosphere. Aesthetic devotion is not incompatible with such naivety, however [...] The immediate emotion of divine atmospheres [in religious devotion] and the indirect emotion in the distancing participation of aesthetic devotion can [...] be coordinated in a labile synthesis." Aesthetic enjoyment is different: "The pleasurable surrender to being affected [...] is aesthetic enjoyment. It presupposes [...] aesthetic devotion, which originally mediates between distance and emotion in the face of the aesthetic image and the atmosphere radiating from it [...]; therefore, it is actually an epiphenomenon in the whole of aesthetic behaviour. [...] As soon as aesthetic enjoyment denies devotion and disregards

aesthetic devotion in the face of a unique appearance: This seems to be what Benjamin has in mind here. Hence the emphasis that he means *these* mountains and *this* twig, that it is about this unique appearance or, as it is said later with reference to an ancient statue of Venus, that it is a question of "this uniqueness, in other words: its aura." (Benjamin 1969, 6).

If Benjamin emphasises the uniqueness of the appearance, the singularity of the aura, then primarily, because he wants to point out that precisely this appearance with exactly the same aura, cannot be repeated, reconstructed, or reproduced. The auratic appearance is not available. It is not the humans who possess it, but so to speak, it possesses them. For: At the moment of the auratic illumination of an object, this auratic appearance seizes the contemplating human, fascinates him, resp. as Benjamin puts it in the Passagenwerk, "takes possession of us".

Just as pertaining to Benjamin: "Aura cannot be mechanically reproduced", also applies the converse: "Mechanical reproduction causes the aura to disappear". It is the authenticity of the object, its fully sensory and holistic presence in reality with which and through which the "aura" can be experienced. As Benjamin puts it: "The here and now of the original constitutes the concept of its authenticity." (Benjamin 1980, 476) The photographed tree, which can be reproduced at will as a photo, is a copy of the original and not the original itself. As Stephan Grätzel phrased it: "Reproductions [...] are primarily images of an original [...] They are oriented towards repeatable data and renounce their own temporal and spatial originality" (Grätzel 1997, 32). Especially with regard to works of art, this means that according to Benjamin "a most sensitive nucleus – namely, its authenticity – is interfered":

> The authenticity of a thing is the essence of all that is transmissible from its beginning, ranging from its substantive duration to its testimony to the history which it has experienced. Since the historical testimony rests on the authenticity, the former, too, is jeopardized by reproduction when substantive duration ceases to matter. And what is really jeopardized when the historical testimony is affected is the authority of the object. One might subsume the eliminated element in the term "aura" and go on to say: that which withers in the age of mechanical reproduction is the aura of the work of art. This is a symptomatic process whose significance points beyond the realm of art. (Benjamin 1969, 4).

---

it, it loses the ground on which distance in emotion becomes possible in the first place, and easily degenerates into unfounded nibbling at stimuli." (Schmitz 1977, 641, 645).

It is now striking that Benjamin's concept of aura consists of four important components, components that often merge into one another or are not clearly separated in his study: *first* the 'grown' *uniqueness* of the object, *second* the sensual *presence* of the original, *third* the aesthetically-devout-contemplative *distance* of the beholder and *fourth* the singular-auratic emotional agitation of the beholder. The first two components primarily concern the object in which aura is experienced, the last two components primarily concern the subject who experience the auratic object. Only these four components, we could say, constitute the full concept of aura. Where only one of these components is missing, the concept of aura is incomplete, deficient: This can be nicely illustrated by an example chosen by Benjamin: a Dadaist collage pieced together from discarded everyday items, from worthless or indecent objects. If you stand in front of such a work of art now, you are dealing with an object that became unique, with a sensually present original; so the fist two components are quite present. However, the last two components do not really manifest themselves: neither an aesthetically devout contemplative nor an auratically poignant state of experiencing can be achieved. And this was also the intention. From the very beginning, the Dadaist artist intended that his work of art be understood not as an object of aesthetic contemplation and auratic experience, but as an affront, shock, disillusionment, disenchantment.

As Benjamin puts it:

> The Dadaists placed much less emphasis on the commercial exploitability of their artworks than on the unexploitability as objects of contemplative immersion. They sought to achieve this unexploitability not least by a fundamental degradation of their material. Their poems are a 'word salad'; they include on scene expressions and every imaginable linguistic scrap. Their paintings, to which they mounted buttons and tickets, were no different. What they achieve with such methods is a ruthless destruction of the aura of their creation by imprinting the brand of reproduction by means of production. It is impossible to allow oneself time for contemplation and statement in before a painting by Arp or a poem by August Stramm, as one would before a painting by Derain or a poem by Rilke. (Benjamin 1980, 501f.)

5.2 *The reception of Benjamin's concept of aura by Gernot Böhme and Michael Hauskeller*

In his volume of essays "Atmosphäre" Gernot Böhme attempts to work out and specify Benjamin's concept of aura against the background of Schmitz' concept of "atmosphere" (see chapter 4 and Schmitz 2023). Aura, according to Böhme, is in the first place "that atmosphere of distance and demanding respect [...], that surrounds artworks." (Böhme 1995, 26). The aura of an artwork cannot be grasped solely by its objective qualities. It goes beyond this, it is a "more" that makes itself noticeable as an atmosphere which envelops the artwork, as it were, spatially outpoured and emotional. Therefore, Böhme concludes: "*Aura,* in a sense, designates the atmosphere in general, the empty characterless envelope of its [the artwork's] presence." (ibid.)

Böhme thus emphasises above all the second component of the concept of aura that I developed above: the component of the sensual presence of the original. The components one and three are also addressed, but remain in the background – for example when Böhme speaks of the "uniqueness of artworks" or of the "elusiveness and distance which can be felt in artworks" (Böhme 1995, 27). Only the fourth component, the singular-auratic emotional agitation of the beholder is denied. For, according to Böhme, every artwork has, by virtue of being an artwork, an aura i. e. atmosphere; every artwork disperses, at least in tendency, an "atmosphere of distance and demanding respect". Even the Dadaists did not entirely succeed in shaking off this atmosphere. "For precisely, by declaring a readymade an artwork, Duchamp gave it an aura, and now his readymades are to be found in museums demanding distance and respect just as much as, say, a sculpture by Veit Stoß. The avant-garde did not succeed in getting rid of the aura". And from this perspective, Böhme, turning against Benjamin can also claim: "The aura is not unique, but repeatable." (Böhme 1995, 26f.).

But his student Michael Hauskeller has now opposed precisely this point: In his essay on "Atmospheres in Nature and Art" Hauskeller asks: "Does a thing acquire an aura (as Benjamin understood this word), only because [...] a recognised artist [...] *declares* it to be art or because it is exhibited in a suitable setting such as a museum? I don't want to rule out that this can happen, but is it necessary? I don't think so." (Hauskeller 1999, 106). We must rather distinguish the "atmosphere of distance and demanding respect" that can basically be attributed to any artwork from the aura which always bears the character of a unique event. And this unique aura is: "Unmistakable, never expected, unrepeatable, singular, unheard of and unconventional" (Hauskeller 1999, 107). By emphasising this unique event character, Hauskeller brings to the fore precisely what I called the fourth component above – the unique auratic emotional agitation of the beholder:

> The encounter with a painting or a statue can, even today, become an event and make us feel both presence and transcendence [...] if we only open ourselves up to what is revealed to us. This encounter is then as unique as Benjamin's summer afternoon contemplation of a twig in the shade of a tree. (Hauskeller 1999, 107).

Now Hauskeller brings to the fore the fourth component of Benjamin's concept of aura, which was denied by Böhme. But he does it so vehemently that other components especially the first two dealing with the uniqueness and the presence of the original almost become trivial. The event of emotional agitation, which Hauskeller tends to view as subjectivist, is so dominant here that the object of this experience, i.e. the original, fades in comparison. Hauskeller is not concerned with the experience of the presence of the *original*, but with the experience of a "pure presence". Therefore, he can claim: "The aura is the experience of presence, and in fact of the pure presence, because it is not this or that which becomes present, although it is something specific, something special, through which we experience presence." (Hauskeller 1999, 104). But now, for his part, Hauskeller overstepped the mark. For in the dominant sphere of pure presence, the question of the object's particularity becomes more or less irrelevant. However, it was precisely this particularity and 'grown' uniqueness of the object that turned out to be so important for Benjamin's concept of aura. Therefore, in the future, we will certainly proceed more appropriately if we do not allow ourselves to be blinded by the concept of pure presence, but address like Gernot Böhme, the particularity of objects in the respective manner of their bodily and sensual presence, in the manner of their emergence from themselves – for which Böhme introduced the expression: the "ecstasy of things" (Böhme 1995, 33, 166-172).

5.3 *On the phenomenon of the aura of a place: the example of the Judengasse ruins*

In the following, I would like to use an example – the ruins of the Judengasse in Frankfurt – to demonstrate how and to what extent aura is not only a philosophical, but also, and above all a lifeworld phenomenon.

First a brief look at the history of Frankfurt's Judengasse. In 1463, the city of Frankfurt had a ghetto built outside the city walls: the so-called Judengasse. The approximately 100 Frankfurt Jews were forced to live there. Over the next 150 years, the ghetto's population rose rapidly. Around 1600 approximately 2,700 Jews lived in the ghetto. All vacant lots were now built on, on each side of the approximately 375 metres long Judengasse stood about 100 buildings, which, due to lack of space, were narrow elongated, tall and cantilevered. The lane was narrow, the lighting conditions were poor, and the living conditions were oppressive. The ghetto gates had to remain closed overnight and on Christian Sundays and holidays. These ghetto conditions persisted until 1811 (Napoleon). Afterwards Frankfurt's Jews were allowed to live freely. However, actual legal equality did not occur until 1864. From the mid-18th century onward, the Judengasse became increasingly dilapidated. Its systematic demolition began in 1874 after the collapse of several houses.

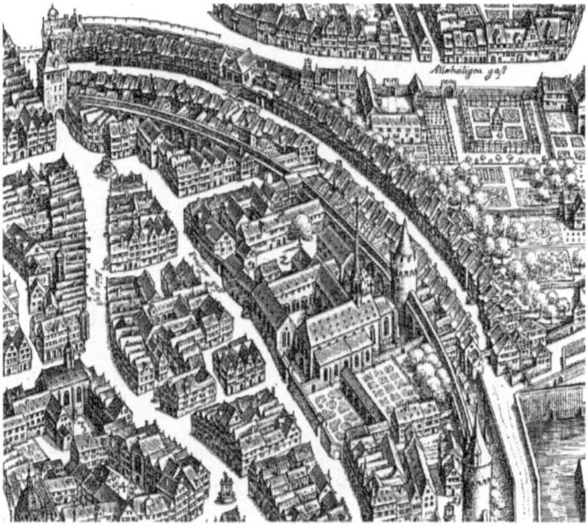

[Pic. 32: Jewish Ghetto Frankfurt. Engraving by M. Merian. Detail, ca 1628. The ghetto is the curved, narrow street outside the city walls.]

For a good hundred years, the Judengasse remained quiet. It was not until 1986 that it returned to the spotlight. For a new administrative building was to be built on its historic site: the customer centre of Frankfurt's public utilities. During the planned excavation work, it was to be expected that historical relics would come to light. Therefore, the monument protection authority intervened, and finally, the Hessian Minister for Science and Art, Mrs. Dr. Vera Rüdiger, imposed on the city of Frankfurt the condition that any finds at the site had to be secured. In 1987 the Judengasse was properly archeologically excavated, examined, and documented. Various house foundations and cellar rooms, including the foundations of an old Jewish ritual bath – a so-called mikveh from the time of the Judengasse's founding – as well as the clay and faience dishes, stucco fragments, and the like, were discovered.

The exposed house foundations and cellars of the Judengasse were so impressive, such an expressive, instructive, shameful, and urging testimony to marginalised, cornered and ghettoised Jewish life, that when it was decided in August 1987 to clear away all the ruins and reconstruct only a small part of them in a museum in the basement of the future customer centre, there were violent public protests and demonstrations. The demonstrator did not want a museum, a reconstruction, or what the called a "history disposal park", but rather the atmospherically charged, expressive, and self-explanatory original. After some of the ruins had been removed, the remaining ruins were occupied. The occupied square was finally cleared by police force (on September 2nd), and the ruins were completely removed. From 1989-92, the Judengasse Museum was built in a basement corner of the municipal utility building. There, roughly at the location of the actual excavation site, the foundations of the houses "White Ram", "Red Ram", "Sparrowhawk, "Stone House", and the two ritual baths or "mikvehs" (the cold bath from the 15th and the warm bath from the 18th century) were reconstructed (cf. Heuberger 1992a, 100-161; Heuberger 1992b, 5, 10-37; Korn 1988, 153-158). But what so many people experienced at the original ruins – the aura of this site – had vanished.

I would now like to ask specifically about this previously so clearly felt, but later vanished, element. I would like to ask: How was the aura of this site perceived? People often talked about the "speaking stones" of the ruins. Ludwig Börne's phrase: "Where the dead are silent, the living stones speak all the louder", became a ubiquitous motto. In his essay "Damnatio memoriae – The Börneplatz as a place of collective forgetting" Dieter Bartetzko recalls the observation "that all opponents of the demolition of the ghetto's remains repeatedly invoke the harrowing aura, the intensity emanating from the sparse walls [...] For all those involved, the stones screamed in a visibly unbearable way." (Heuberger 1992a, 29). Another person of those involved, Claudia Michels, had a very similar experience: she, too, speaks of the screaming stones of the Jewish ghetto (Heuberger 1992a, 61). And the writer Eva Demski says: "the beautiful phrase 'speaking stones' has made the rounds. But then they spoke, too. But how can someone who does not listen to people understand stones"? (Heuberger 1992a, 159). Less metaphorical, but still pointing to a similar point, Thomas

Kirn writes: "In the middle of present day Frankfurt, the relics of an urban culture literally lay open, visible to everyone." (Heuberger 1992a, 59f.).

Salomon Korn describes most impressively and nuancedly what could be experienced at the ruins of the Judengasse and what was irretrievably lost there:

> One didn't have to be a historian, or possess detailed historical prior knowledge, to be able to relive the oppressive narrowness of the houses and lanes in the excavated cellar remains. The desolate misery was physically palpable at the excavation site – not least through the randomly formed and destroyed remains of the walls with their partly bizarre silhouettes [...] Anyone who wasn't numb could experience intuitions [...] of inescapability and the focal narrowing of life's horizons. What a wealth of dark images was suddenly able to emerge from these stones up close and bodily appreciable! The stones have been (meanwhile) removed. Now, five of the removed cellars, [...] within the planned customer centre, are to be reassembled from original stones and additions, and their historical significance made understandable through museographic and didactic processing. As necessary and indispensable as such accompanying explanations are, nothing will be felt of the narrowness, nothing of the aura of the authentic site, and certainly nothing of the aura of the original. Original sensuality cannot be replaced by even the best information and neither can do so reprocessed ruins and mock-ups of destruction. As they are no longer able to trigger a sense of 'recognising fright' in the viewer like the ruined originals, they run the risk of becoming neat and well-commented wall remains on the edge of embarrassment. Despite images and videos, the museographic and didactic presentation is aimed first and foremost at reason and discursive logic. It cannot convey the genuine experiencing, as was possible with the uncovered remains of the Judengasse. (Korn 1988, 159f.)

Of course, the question arises whether Salomon Korn adopted the term aura from Benjamin, with which he tries to describe the "genuine experiencing" of the "speaking stones" of the Jewish Lane ruins. I think so. It is very likely that Korn, who studied architecture and sociology at the Technical University of Berlin and the Technical University of Darmstadt, was familiar with the Benjamin text in question. Another of those involved in the Judengasse case, Georg Heuberger, the later director of the Judengasse Museum, was certainly familiar with this, for he refers directly to Benjamin's concept of 'aura':

> The public debate surrounding the architectural remains of the Judengasse has undoubtedly made it very clear that the ruins, as authentic and very vivid testimonies to Jewish history in Frankfurt, exert a great fascination. One could paraphrase this with Benjamin's concept of 'aura'. Thus, the ruins convey

> to many observers the impressive experience of the tangible and sensual proximity of history, combined with the experience of its temporal distance and strangeness, which eludes immediate comprehension. In this case 'distance' arises from the feeling of almost sacred inviolability that is often shown appreciation for architectural testimonies to Jewish history in Germany after Auschwitz. (Heuberger 1988, 257f.)

But it is not so crucial here which of the participants were familiar with Benjamin's concept of aura and which were not. What is crucial that most of the participants repeatedly addressed, admittedly with varying degrees of emphasis, the four components – *first* the 'grown' uniqueness of the object, *second* the sensual presence of the original, *third* the aesthetically-devout-contemplative distance of the beholder and *fourth* the singular-auratic emotional agitation of the beholder. And these components were not addressed as intellectually-historical or humanistic topoi, but as experience and lived "life spatial" phenomena, as "life spatial" reality; as a reality that cannot be scientifically reconstructed and technical reproduced, but certainly be lived and experienced humanly.[7]

5.4 *The aura of a place as atmosphere, expressive image, "archetype", tendency to movement and forgathering potentiality*

Finally, I would like to take another look at the repeatedly appearing phrase of the "speaking stones". The discourse of the "speaking stones" is quite obviously a metaphorical one. It was not meant that the stones spoke in human language, but, as Eva Demski put it, the stones were "silent but unmistakable" (Demski 1988, 18). My question now is: How could one express what is meant less metaphorically? First, as already done, with Benjamin's concept of aura. But some things can be grasped more clearly and with a phenomenological foundation, not least against the background of the insights in my monograph "The Spirit of a Place" (Kozljanič 2004a, 2004b).

The already quoted Dieter Bartetzko spoke of a "shattering aura" and of an "intensity" of the place. And Eva Demski spoke of the "historical richness of an atmosphere" at this site, and how those "who visited this site were overcome by a somewhat different feeling than one has in a museum,

7 On the reality of „lived space" and the concept of lived and experienced space, see: Bollnow (2020, 33-43); but above all: Dürckheim (2005); and: Hasse (2006, 165-167; section "Heimat als 'gelebter Raum'").

where history is detoxified and tailored with a catalogue and cafeteria." (Demski 1988, 17f.). Hence, it would be appropriate to describe what is meant less metaphorically with the concept of atmosphere developed by Schmitz; or more precisely, with the concept of a *spatially anchored and object-bound atmosphere* developed by Schmitz and concretised by me (cf. chapter 4.3 and Kozljanič 2004b, 330-335).

However, it was not only the site-anchored and intense atmosphere that was important here, but also the impressively expressive image of this site (cf. Kozljanič 2004b, 263-287). Not only the purely atmospheric appealed to and moved the people, but also the visible expressive image, so to speak, the physiognomy of the exposed Judengasse foundations, the curved nature of the lane, the restriction and density of the visible floor plans, the cellar corridors, the flights of stairs and vault remains. In this spirit, Eva Demski says: "The site was there, [...] it presented itself [...]. You had to see the depth of the lane at night, on those very quiet nights." And a little later, she speaks of the "delicate and mighty stone skeleton of the lane, slightly curved like a fish", which you "could clearly see ", that "was really there" (Demski 1988, 19f.).

According to Klages the perception of the image can be deepened and intensified. If you surrender to the inherent dynamic of impressively expressive images – or to use Böhme's words: If you allow yourself to be radically grasped and carried along by the "ecstasy of a thing" – it can happen (according to Klages) that intuitions (in extreme cases even visions) are induced: intuitions of affectively charged, profoundly meaningful images with, as it were, archetypal potentiality. I then speak of "image intuition". In such a case Klages speaks of "primal images" and "image vision". I presented and modified Klages' theory of "primal images" in my genius loci monograph (Kozljanič 2004b, 288-300).

Salomon Korn described in his above quotation quite a comparable process of image intuition:

> Anyone who wasn't numb, could experience [when contemplating the expressive image of the Judengasse] intuitions [...] of inescapability and the focal narrowing of a life's horizons. What a wealth of dark images was suddenly able to emerge from these stones up close and bodily appreciable! (Korn 1988, 159f.)

The comparison used by Eva Demski in the penultimate paragraph – that the "stone skeleton" of the Judengasse lay there like a slightly curved fish – not only indicates that physiognomic perception is directed towards the expressive image, not only that the expressive image can, spontaneously, elicit and provoke deeper images (in this case, the metaphor of the fish-like, curved stone skeleton), it also reveals another important feature of lived and experienced space. It is the feature of "kinetic suggestions", which comes from the spatial phenomenon causing the contemplator to participate in this movement internally and may be externally as well. Although the stone skeleton lay there, so to speak immobile and static, it nevertheless possessed a motive tendency; this was due to the mere fact that the contemplator's eye moved along it, nestled against it, and thus, admittedly just subtly, fulfilled an emotional "quiet" movement. The curved, fish-like gesture of the stone skeleton thereby (but by no means only thereby) triggered a suggestion, a "kinetic suggestion", as Hermann Schmitz aptly called and described this phenomenon. The kinetic suggestion is not a purely subjective process. It is closely connected with the progression of the shape of the object (resp. its expressive image). Progressions of the shape and movement suggestions, according to Schmitz,

> are to be equally found [...] in shapes that are neither movement nor in motion. Progressions of the shape and kinetic suggestions altogether do not need to be felt [in terms of empathised] or otherwise added to what we call perception, but they rather are [...] immediately perceptible. (Schmitz 1978, 41, cf. also 38-46; and Schmitz 1995, 141-143, 282-284).

This is why we can state precisely and summarise: Auratic sites are endowed with a unique quality of appearance, a unique *character*: And based on this character we can distinguish (in retrospect, and often more poorly than well): *atmosphere*, *expressive image, primal image* and indeed *kinetic suggestion (or movement tendency)*. But that's not all. For there is another important characteristic: the *gathering potentiality*.

To comprehend this last characteristic more clearly, one could invoke Heidegger's idea (presented in chapter 2.6; and in my monograph, Kozljanič 2004b, 301-320) of the gathering potentiality of a place (admittedly weakened, less focused on the 'Sein' and more on the historical and social aspect, thus almost too allegorically interpreted).[8] For the Judengasse ruins had a gathering quality – in many respects: They gathered and bundled spatial relics and historical memories, gathered and bundled the most diverse people. A place of encounter, where past and present spaces and times touched and touched each other, a place of encounter where people of most diverse worldviews, ages and backgrounds met, a place where

8 As presented in chapter 2.6; and in my monograph, Kozljanič (2004b, 301-320).

spatial, historical, and social elements could converge. As Eva Demski so aptly put it:

> What this place could have become if it had been left alone! The trite alibi phrase of a meeting place, which easily brings to mind empty speeches and wreath ribbons, took on its own, ne quality. There it was and the stones were not just stones but, but occasions for questions and answers." This was not about "self-celebration architecture", but "about the only authentic place that could make clear how Jews were treated. Despite the many people who were often there, a few steps away you could be completely alone and think back. Perhaps that was precisely what they [the advocates of the demolition of the Judengasse] did not want: a place that is so quiet and so disturbing. (Demski 1988, 19).

5.5 *On the socio-historical dimension of an auratic site*

The previous text (apart from the second section referring to Gernot Böhme and Michael Hauskeller) was the basis of the lecture, which I gave on 2nd April, 2010 as part of the 49th annual conference of "Symbolon – Gesellschaft für wissenschaftliche Symbolforschung" (Society for Scientific Symbol Research) in the Judengasse Museum Frankfurt. In the subsequent discussion, several questions were raised, of which I want to answer a particularly important one, in a similar way, but in greater detail than at that time. One listener expressed fundamental reservations. Her basic objection was to ask whether the social and socio-scientific (and thus probably also the historical) dimension had not been completely underrepresented resp. had not been taken into account in my lecture.

I answered: Yes. But this was because the main question of my lecture was not the psycho-social and socio-historical constituents, but about the phenomenological description and explication of the aesthetically-sensory (aura) experience made by the individuals on this site. But, it would be of course necessary and sensible to examine the specifically social dimension of the Judengasse conflict separately.

One could, for example, ask here about the role of socio-cultural memory, more pointedly: One could ask whether the atmosphere experienced by those involved in the Judengasse ruins was not, above all, a socially constituted (or even socially constructed) atmosphere – and not a purely aesthetically-sensory, spatially anchored, and object-bound one. We could, indeed we must, ask this question if we want to develop a deeper understanding of the Judengasse conflict. However, we should not overlook that in the cited empirical reports both dimensions – the phenomenologically-aesthetic and the socio-culturally constitutive one – are inextricably interwoven. In my presentation, I primarily concentrated on the *object-bound* aesthetic experience, because it is neglected in most theoretical debates on such a topic. I did not specifically address the subject- and society-bound dimension of the socio-cultural memory, but – tacitly – assumed. Only at the very end, however, did I again point to the social component from the object or site side: the socio-culturally gathering potentiality of the site.

And in this respect, the listener's critical question is correct in a first appropriate way: I could resp. should have emphasized and differentiated the socio-cultural dimension of experience more strongly. In this context, however I can refer to another publication, where I elaborate on this very important point. Yet, even in this publication, I was not so strict and absolute to separate the aesthetic dimension from the socio-cultural dimension of "communicative", "collective", and "cultural memory" (Maurice Halbwachs, Jan Assmann), but addressed both dimensions in their mutual relationship, convinced that they normally appear inextricably intertwined in lifeworld experience (Kozljanič 2010b, 157-176).

A completely different question, of course, would be whether the atmosphere essentially (co-)conditioned by the socio-cultural memory is a socially constructed phenomenon or more precisely: a *social construct*. For my part, I use the term "construct" very cautiously, because it presupposes a designer or draughtsman. And it presupposes that a draughtsman consciously and intentionally assembles a construct from various elements; much like constructing a machine. But all of this does not apply, or at least marginally, to *the* socio-cultural phenomena that spontaneously form and/or develop and change (multi-casually conditioned and multi-functionally oriented) on the basis of socio-cultural memory. I would therefore only allow this term to apply to those phenomena that were consciously and intentionally brought about by social designers (first and foremost, of course, all kinds of social cultural policy planners, but also planners involved in infrastructural measures; in a more extreme sense, however, also political ideologists, media manipulators and other demagogues). Or to put it in another way: The fact that something became social and was socially constituted does not mean that it is necessarily socially constructed. We should distinguish here between social constitution and social construction.

Back to the basic objection raised: It is not only correct in that I should have emphasized and differentiated the socio-cultural dimension more strongly, but also in that I should have taken the historical dimension more into account. This is the *second way* why this objection is justified. What is at issue here could be summarised under the term "historical testimony". And the objection could be specified: The historical testimony did play a central role in the case of the Judengasse conflict. And this testimony is a difficult point in my definition of aura, which is based on Benjamin. For it could be no longer phenomenologically justified, and it does not simply appear (as in Benjamin's first example) on a summer afternoon, lying on one's back be it close up or from a distance. From this perspective, one could possibly object to Bartetzko and Demski: The – physical and politically legal – confinement of the Jews in Frankfurt before equalisation cannot be sensed or recognised from the foundations. Foundation walls of this type of building are always somehow "narrow", even non-Jewish ones. If there is something auratic about the cellars, it is through their historical testimony. These walls, simple cellar walls, cannot – as Demski would have us believe – "make clear [...] how Jews were treated". However, they could, through their historical authenticity, trigger feelings of concern, but only against the background knowledge about the Judengasse.

I would also like to respond to these concerns: The first of the components I identified is: "the 'grown' uniqueness of the object". The word 'grown', which refers to the "uniqueness of the object", also contains in nuce what one could call "historical testimony". The uniqueness of the Judengasse ruins lies precisely in their historical 'grown' materialisation. And this materialisation implies not only that those cellar walls were once built, later filled in and built over, but also that these walls and foundations therefore were built so tightly because the Jews had been forced into a ghetto (which had become far too cramped by 1600 at the latest). This materialisation also implies that Jews in Frankfurt were ghettoised in the first place from 1463 to 1811. It also implies the further history up to the National Socialist mass murder of the Jews and the German post-war "inability to mourn" (Margarete and Alexander Mitscherlich). Or to put it in another way: It was not just any lane whose foundations were exposed, but the *Judengasse* (Jewish Lane). And it was not any city in which the conflict ignited, but a German city and, what is more, a German city that is and was particularly closely entangled with Jewish history and fate: Frankfurt. And it was not just anytime when the protests took place, but the 1980s, with all their protest movements and civil initiatives, with their social, ecological and politico-peace commitment.

What I said in chapter 3 with regard to urban genius loci sites, also fully applies to the Judengasse ruins in their "materialisation".

Such sites "are not only historical but also layered places. They are layered places in two ways: firstly, historical relics deposited and sedimented: rubble, waste, everyday objects, bones; secondly, memories accumulated in the social memory of the residents and/or those who transmit history (in written or oral form). Both forms of sedimentation overlap and mingle in the life-world phenomenon of the site (cf. Freyer 1990). Told and remembered stories attach themselves to relics sedimented in the ground; sedimented relics themselves tell or initiate stories, evoke memories. Genius-loci-related urban planning should take these sediments into account: for only sites with sediments can be multi-layered; but being multi-layered means: having rich depth; having an 'aura'; having an aura means: evoking the numinous.)"

The component of "materialised uniqueness" of Frankfurt's Judengasse thus implies the historically stratified character and implies its historical testimony and also implies what was deposited in the socio-cultural memory.

To avoid being misunderstood: I do not intend to refute the objection in question with my remarks. This objection hits on an absolutely central point. And I would therefore like to add a fifth component to the four components I extracted from Benjamin's concept of aura. I would like to explicate what was merely implied in one component of the aura concept with a new component. I propose to call it the "*multi-layered-remembering testimony of an object*" and setting it to one side the four existing components (the 'grown' uniqueness of the object, the sensual presence of the original, the aesthetically-devout-contemplative distance of the beholder, the singular-auratic emotional agitation of the beholder).

However, I would like to contradict the following objection mentioned above: that the cramped conditions of ghettoised Jewish life could not be perceived in the ruins, and that all types of foundation walls (Jewish or non-Jewish) were also somehow narrow. Of course: Foundation and cellar walls usually have something "clammy", "shaft-and cave-like" and in this sense also something "somehow narrow". But in the case of the Jewish streets, the narrowness was not "somehow narrow" but particularly narrow. The extremely narrow, crowded, elongated floor plans of the houses were quite obviously different from the floor plans of other comparable non-Jewish houses.[9] And this particular *physical narrowness* was clearly noticeable on site. The *politically-legal* narrowness, however, could not be determined without remembered pre-knowledge of the exposed foundations. It only became apparent to those who knew about the history

9 See the photo of the excavated ruins in: Heuberger, G. (1992b, 92); but also the city views and property plans from 1600, 1628 and 1711 (ibid. 19, 15 u. 28).

of the Judengasse. It was only existing to those who remembered the socio-historical significance closely linked to the Judengasse: But then it was existing to them; existing not as a purely internally subjective idea, but as a spatially anchored and object-bound "shattering emanation" (Bartetzko), as a "wealth of historical atmosphere" (Demski) as a "wealth of dark images" and memories that could suddenly unfold from these stones "very close and bodily experienceable" (Korn).

Here we have obviously reached a point where the paths of the psychological and phenomenological approaches diverge. Psychologically speaking, this is a projection process: The remembered meaning is brought along by the subject and then projected onto the site. Phenomenologically speaking, however, it is a process of actualisation and articulation. The remembered meaning is released from the site then actualised and articulated by the subject. The meaning *seems* to be somehow and vitally connected with the site. It seems: And in this case, if we want to recognise the phenomenon experienced in the lifeworld as such, we have no right and no possibility of absolutely separating it from the site and ascribing it to something else. If we dissolve the appearance here, we dissolve the phenomenon likewise. This does not mean, however, that we should be content with simply superficial appearances. After all: We can (and should) examine appearances in depth. Apparent phenomena have not only surface, but often depth, too, especially when they are *phenomena so eminently symbolic*, as in the case of the Judengasse ruins.

5.6 *The Judengasse ruins as an auratic and symbolic site*

This brings me to a final point; a point that was also implied in my lecture, but not specifically explicated: namely the question of the Judengasse ruins as a symbolic site, as an *auratic-symbolic site*.

A symbolic site is a place that functions as a symbol in a special way, that symbolises something and, in this sense, is a symbol itself. And because the site is a symbol, it can be, of course, considered from a symbolically theoretical perspective, for example with Ernst Cassirer's "theory of symbolic forms":

> A 'symbolic form' is to be understood as any energy of the spirit through which a spiritual meaning is linked to a concrete sensory sign and inwardly appropriated to this sign. In this sense, language, the mythically-religious world and art, in each case, appear to us in symbolic forms. For in all of them there is this basic phenomenon that our consciousness does not content itself with

> receiving an external impression, but it associates and gives essential features to every impression with a free, expressive activity [i.e. sign]. What we call the objective reality of things is confronted with a world of self-created signs and images, which assert themselves against it in independent fullness and original power. (Cassirer 1994a, 175f.)

Accordingly we could see in a symbolic site a sign created by humans, a spiritual representation that corresponds to a specific form of representation – Cassirer says: "symbolic form". Symbolic forms are, just listed by Cassirer: language, myth, and art. Science, which is missing from this list, represents the fourth symbolic form for Cassirer (1994b, 18).

Interestingly, however, a symbolic place, like the Judengasse, can only be assigned rather imprecisely to Cassirer's symbolic forms: It subverts art in the direction of myth, however, without becoming absorbed in the mythical and religious. Yet despite an accurate definition of the most general characteristics of the sign and of symbolic linking and world-conception, a deficiency is revealed in another aspect as well. If the concepts of sign and symbol tend to be used synonymously, as with Cassirer, and if symbols and signs are seen only as anthropogenic carriers of meaning and representations, then the irreplaceable depth of meaning and significance of the symbol concept, which extends into the auratic and numinous, is thereby also levelled out – together with its inherent moments of unavailability. This loss of differentiation was therefore rightly criticised by life-philosophers against Cassirer.

Otto Friedrich Bollnow, for example, asks

> whether it was appropriate to describe these general forms as symbols; for this suggests confusion with other linguistic usage, as common in theology, where the word 'symbol' is used in a narrower sense, namely as a double-meaning (or multi-meaning) sign, so that a deeper meaning is hidden behind the literal, obvious meaning. For example, washing with pure water symbolises moral purification. (Bollnow 1980, 2).

However, there is an even more serious reason why Cassirer's equation of symbol and sign ultimately fails to reach the dimension of the auratic concept of symbol (which is in question in the case of the Judengasse ruins). This reason is predetermined by Cassirer's neo-Kantian, transcendentally-philosophical view that direct experience (outside the human sphere of symbolic representation) is not possible for humans, since it always appears to them only through the medium of signs and symbolic forms as *mediated* – and thus no longer immediate reality. According to Cassirer, reality and experience are only possible on the (mediating and mediated) 'stage of the spirit'. According to Cassirer, a human is an *animal symbolicum*: "He lives so entirely in linguistic forms, in artworks, in mythical symbols or religious rites that he cannot experience or perceive anything except through the interposition of these artificial media." (Cassirer 1960, 39).

The question at issue here is no less important than that what can be experienced in and through symbols transcends the sphere of the purely human, purely social, purely culturally mediated and cognitive and can lead into areas of more immediate experiencing, sensing and experience. It is the question of the immediacy of life that ultimately divides opinions. Cassirer pursued this question intensively, as the posthumous manuscripts from 1928 and 1940 show, which were edited in 1995 (and were to form the unpublished 4th volume of his "Philosophy of Symbolic Forms"). During this time, Cassirer developed, tested and sharpened his basic concept of transcendental philosophy through symbol-theoretical appropriation and criticism of life-philosophical positions (Friedrich Nietzsche, Ludwig Klages, Henri Bergson, Oswald Spengler, Georg Simmel, Max Scheler).

Christian Möckel, in his study "Das Urphänomen des Lebens – Ernst Cassirer's 'Philosophy of Life'" (Möckel 2005) traced this debate with the philosophy of life. As Möckel explains, Cassirer's basic concept corresponds in essential points to Georg Simmel's life-process / cultural-form dialectic. However, with one crucial modification: For Cassirer, the poles of "life" and "form" (or "idea") are not the original, but rather their correlation tangible in the symbolic medium and as a symbolic medium. Where Simmel ultimately remains undecided on this question, Cassirer takes a transcendental-philosophical side: for the symbolic medium. As Möckel (somewhat circularly) puts it: "The poles of human life, understood from the perspective of the symbolic medium as a correlative relation between 'life and idea', have their primacy in the correlation: It is precisely this correlation that is the primary certain an the primary given, not the antinomy." (Möckel 2005, 245). As Möckel shows, Cassirer attempts to unmask the life-philosophical discourse of '*pure* (experiential) life-immediacy' or the immediate, pure vision of life as untenable. And it is indeed questionable whether we will ever be able to justify theoretically such *pure* and *absolute* immediacy. But the life-worldly experiential value remains that there are more mediated and less mediated, more original and less original experiences; and thus the bone of contention.

But what would happen if we were to delete the little word 'pure' from the life-philosophical discourse of '*pure* (experiential) life-immediacy' (Lebensunmittelbarkeit)? What would happen if we were to acknowledge the life-world distinction between more original and less original experiences without immediately hypostatising the *one* original experience *as such*? If we were first to consider what the concept of (experiential) life-immediacy actually refers to: namely the abundance of comparatively immediate, involuntary, spontaneous sensory and bodily experiences? What would happen then? Then the concept of (experiential) life-immediacy would be transformed. Strictly speaking, it would no longer be a static, thing- ontological or even metaphysical concept of substance, but a dynamic, process-oriented (genesiological) concept – life-world concept of experience. As such a concept of experience, it would refer to life experiences, which certainly allow for gradual differences: differences of the type "more mediated" – "less mediated", "immediate" – "more immediate", "intense" – "more intense", etc. We could then (and most philosophers of life did so, even if they repeatedly slipped or were pushed aside in the theoretical battle towards the concept of substance) increasingly allow and illuminate the (experiential) life-immediacy guided by this concept of experience. This would then be a project of enlightenment, or better of illumination, which would not have to shy away from immediacies, powerful feelings and distresses, but could engage with them – not blindly, but with a life-philosophically thoughtful approach, not obscuring, but illuminating.

But Cassirer ultimately wanted to see this engagement with the (experiential) life-immediacy only as a "belief in natural necessity and inevitability". Thus, the moment of freedom contained therein escaped him – a moment of *sovereign* "freedom to", not of *autonomous* "freedom from". And that is why Möckel summarises,

> he fought this belief throughout his life as the 'darkness of irrationality'. Only by understanding and accepting himself as the creator of his cultural, symbolic world of meaning, and living actively within it as a self-aware ethical being, can he escape the contingency and finitude of his vegetative existence, to which Heidegger's existential philosophy once again surrenders him. Every rehabilitation of the immediacy of life (experiencing) favours these irrational moods and obscures the value of rational, symbolic cultural forms as the decisive means of bringing about humanity's self-liberation from the fateful condition of natural existence. (Möckel 2005, 394)

The question remains whether such exclusiveness options are really compelling in this way. Whether, in other words, an overly strong pathic involvement with "real-life emotions" *must* inevitably lead to a destructive "darkness of irrationality". And whether, as Cassirer demands, "its emotional disposition *has to be* refuted", so that freedom and the light of reason are to flourish there (cf. ibid.). There are completely different concepts here. Concepts that assume that freedom and reason do not necessarily have to presuppose the weakening and suppression of sensuality and emotionality, but can blossom just as well, or even more beautifully on the ground of free sensuality and emotionality.

Ernesto Grassi (coming from Heidegger) offered an interesting alternative to Cassirer here when he demonstrated that every act of symbolic formation (in Grassi's words: every légein- resp. metaphérein-act) is not only an act, but also pathos. And that in those acts of formation, that, according to Grassi, are "original" or we should better say "more original" the pathos side is dominant: in the sense that something is suffered originally, i.e. inevitably, and the act of formation which responds, forced in this manner, to the suffered claim, is in this sense more original than an act of formation which arises from other, less "existentially" compelled motives.[10]

Here it becomes clear that Cassirer's transcendental-philosophical view and his concept of symbol is not the only valid one. It can only maintain its claim to validity as long as it succeeds in methodically excluding the life experience underlying all mental representations and cognitions and the more original emotions associated with it (Grassi speaks here of "original passions"). If not, it finds itself confronted with the dimension of experiencing to which Grassi pointed throughout his life: the dimension of being addressed by the original, the dimension of inescapable compulsion, of which Grassi said: "This, like everything first and original, is not inferable, communicable by nothing else; it simply comes *over us*, we sustain it." (Grassi 1940, 47).

Cassirer's definition of the concept of symbol thus means not only a loss of differentiation, but due to the transcendental-philosophical and methodical self-restriction a loss of world-content above all. It overlooks the most important aspect of the symbol: the indicative aspect; more precisely, the aspect, which points to deeper dimensions of reality; even more precisely: the aspect that points to original experiences yet to be had. As Bollnow puts it: "The symbol leads us into a depth of the world which we can basically only grasp by its guiding thread." (Bollnow 1980, 3).

I now state: every genuine symbol *can*, firstly, have a depth of meaning and significance that extends to the auratic and numinous; secondly, it *can*

10 For more details, see: Kozljanič (2003, 27-48: section "The logos as act and pathos: Grassi's theory of légein (1939-70)"; and 48-59: section: "The logos as a fantastic and metaphorical process: Grassi's theory of metaphérein (1970-91)").

extend beyond the sphere of the purely human, the purely social, the purely spiritual towards more immediate experiences; and therefore, thirdly, *can* point to, lead to and open up deeper layers of reality and experience. I would like to draw attention to a fourth aspect: namely the exonerating (evoking) and connecting function of the symbol.

This aspect, combined with the other three, was particularly emphasised by Henri Lefebvre. In his "Critique of Everyday Life" he speaks of the "inexhaustible character" of symbols, which is more "than what discourse and representations elaborate", which rather amounts to a "dark and almost magical participation" and is therefore clearly to be distinguished from the sign in the narrower, semiotic sense. He illustrates this with an everyday example:

> When I say of someone: 'He is cunning as a fox', I attribute to him the qualities that (I think) the fox possesses. Of course, I know that a human being 'is' not a fox, and I also know that nowhere does there exist a platonic idea of cunning, which, among animals foxes and among humans certain individuals share. Nevertheless, I remain in the foreground of these clear, all-to-clear ideas, the clarification of which reveals their logical incoherence. On an affective level, which, – *despite everything that separates me from it* – is close to the childlike (fairy tales and fables), as well as to the archaic (myths) and the imaginary (unheard stories), I say that this person has something common with the fox – and in fact something so real that I behave towards him as to a cunning and vicious animal, dangerous to my possessions and to me, and I strongly advise others (the people to whom I speak) to behave in the same way. The symbol [...] partially *identifies* the two creatures. It demonstrates a *participation* of one in the other. It *evokes* the one through the other and vice versa. The comparison is different when both are viewed by their outward appearance uniting them only through comparative relations like ('so...like', 'more...than'). Finally, as for the analogy pursued for the purpose of reasoning, it highlights the differences. Or it dissolves into a symbolism that gives direction. [...] Every symbol [...] intervenes in the discourse like an exclamation: 'The earth!...The sun!... The night!...' It draws behind it a whole trail of affects and imaginations. [...] It is important [...] to emphasise the particularity of the symbol: that it is inexhaustible and that it has a direct emotional (affective) effect. (Lefebvre 1987, 542f.).

Now we can also understand how and to what extent the above-mentioned historical multi-layered complexity and depth of the Judengasse ruins also correspond to a complexity and depth on a symbolic level. Thus the exposed, closely packed remains of cellars and houses could also be experienced as symbols of excluded and marginalised, oppressed and suppressed late medieval Jewish life and of the modern era. At the same time, the overcoming of such structures of oppression and exclusion (achieved through secular enlightenment and freedom movements) could be experienced: that, for example the Jewish ghetto could be a thing of the past from 1874 (when its demolition began). A modern layer was superimposed over these two older layers: the memory of the Nazi atrocities, the persecution and extermination of the Jews in the Third Reich – and the associated historical responsibility of the citizens of the Federal Republic of Germany: The Judengasse ruins were also a symbol of this, and this too could be experienced at this symbolic site. And right on top, lay the most current layer: It stood as a symbol of how, even in Frankfurt of the 1980s, it was possible to treat such an authentic and auratic testimony to Jewish history so irreverently: The shock that, against the background of centuries of defamation and oppression of Jews, that even against the background of the disaster of the Holocaust – and under today's democratic and history-conscious auspices – such insensitivity could be exerted on the auratic relics of Jewish life in Germany.

What is true of symbols (in an unabridged understanding) is – that they are world encompassing, vivid images of meaning with a pointing gesture; they (can) point to more existential and original layers of experiencing and reality – can now be expressed in relation to the symbolic site of the Judengasse as follows: there moments of past Jewish life and suffering can be experienced (relatively) directly without the interposition of museum processing and mediation – only through social interactions with the site and through the site. The site could thus appeal emotionally in a (relatively) direct way, affect people, and spark political engagement. It could evoke deeper images and layers of meaning, bring to light elements of fate (relatively) directly and vividly; and precisely in this way transcend the sphere of purely human, purely social, purely spiritual – as something humanly, socially, and spiritually available. For it was always clear: Without the original site, it cannot work. And this original site with its aura and symbolic power – my contribution has, I hope, clearly illuminated form various perspectives – cannot be disposed of arbitrarily. Such a site can neither be reproduced nor represented, it can only be experienced in all its multi-layered and multi-meaningful presence.

6.
Current Phenomenological Debate
Schmitz, Seamon, Kozljanič, Griffero

In the last 25 years, the phenomenon of the genius loci has been more accurately described. Gernot Böhme, Tonino Griffero, Robert Josef Kozljanič, Juhani Pallasmaa, Edward Relph, David Seamon, Christopher Tilley and Tomáš Valena should be mentioned here. These authors have one thing in common: they take pre-theoretical lifeworld experiences seriously and use a new phenomenological approach in which the concept of lived space, felt body and spatially manifest atmospheres is important or even crucial. In doing so, they reactualise and revise older phenomenological and life-philosophical theories by Ludwig Klages, Karlfried von Dürckheim, Martin Heidegger, Maurice Merleau-Ponty, Otto Friedrich Bollnow, Christian Norberg-Schulz – but above all Hermann Schmitz.

6.1 *Hermann Schmitz and his concept of situations and atmospheres*

Schmitz criticises the leading philosophers and scientists for cutting off human thinking from the most important parts of spontaneous life experience through exaggerated-abstract constructions out of touch with everyday life. In contrast, Schmitz shows that genuine life experiences not necessarily require or presuppose abstract superconcepts nor singled-out *hard facts* (Schmitz: "objective facts"), but firstly holistic and meaningful *situations*. According to Schmitz, these situations are the original objects of perception, the original phenomena (cf. Schmitz 2019, 73-75; and Großheim 2018). They appear as bodily felt *interwoven ambiance-ensembles* with internally diffuse but rich structure.[1] Although they occur only in a subjective state of affective involvement, they are more than subjective: they are intersubjective *gentle facts* (Schmitz: "subjective facts") concerning reality. It is the task of the phenomenologist, to examine, distinguish and explicate these initially diffuse ensembles. According to Schmitz, poetry is a "gentle explication of

1 With regard to the term "ambiance", cf. Thibaud (2022).

situations" (Schmitz 1995, 461f.).[2] The phenomenologist should also gently verbalise and explicate situations – but, as far as possible, systematically and unmetaphorically (cf. Schmitz 2019).

This raises the question of whether and to what extent phenomenological description and hermeneutic explication should use metaphorical language, or whether it should avoid or even eliminate metaphorical language as far as possible. I have always opted for metaphorical language, but also for philosophically differentiated language. I therefore agree with Christopher Tilley when he says: We should "exploit the tropic or metaphoric nature of language, avoiding a deadened and deadening [scholastic] literalism, to make writing a voice for the stones, the places and the landscapes in which we are bodily immersed." (Tilley 2004, 28).

Be that as it may, Schmitz was able to prove in a conceptually differentiated way that and to what extent genius loci sites are internally meaningful, diffuse and manifold *situations*. There are two main types of situations: 1. Impressive-present situations; suddenly and significantly appearing in an overall impression of felt presence. 2. Segmented-complex situations; appearing fragmented and uncomplete in one or few segments and therefore requiring further knowledgeable additions and contextualisation. Situations are impressive-present "when they suddenly come to the fore in their integrated meaning (like dangerous situations that have to be grasped holistically and answered aptly at a stroke in order not to fall victim to the danger), segmented[-complex] when this is never the case." (Schmitz 2003, 91). I argued above: So while natural genius loci sites largely speak for themselves, at least atmospherically and physiognomically, cultural sites require fuller explanation and/or instruction. Although Schmitz did not draw this conclusion: we may draw it. We can now state why this is so: because natural sites correspond more to impressive situations, cultural sites more to segmented ones.

A specific feature of these subjectively experienceable and intersubjectively communicable situations is that they are atmospherically charged and only become tangible and articulate in their spatially effused 'flowing' atmospheres. Schmitz said of the Loreley and other siren-like spirits of place that they draw their "power of suggestion from the nimbus of a powerful atmosphere", from a "highly emotive" mood (Schmitz 1977, 150).

2 Cf. Alexander Popes famous translation of a Latin verse: "Nymph of the grot, these springs I keep, / And to the murmurs of these waters sleep; / Ah, spare my slumbers, gently tread the cave! / And drink in silence, or in silence lave!" – A beautiful metaphor from 1725: one should tread (and treat) a genius loci site gently!

[Pic. 33: Lurlei or Loreley. Postcard, ca 1900.]

These atmospheres are thus situationally-intersubjectively accessible and subjectively-bodily perceptible, especially in a state of affective involvement. Only then they show themselves as spatially given, i. e. actually lying in the landscape and not (as is often assumed) psychically projected or even socio-culturally constructed. Schmitz usually emphasises these atmospheres as quasi-climatic, fragrance-like, and spatially effused phenomena. This also applies to what Schmitz calls "local divine atmospheres": and thus especially to all numinous genius loci sites (ibid. 128-134; cf. also chapter 4).

6.2 *David Seamon and his concept of genius loci as a lifeworld common presence*

The ambiguous term "sense of place" – used by phenomenologists such as Edward Relph and David Seamon – comes quite close to this concept of situationally accessible and bodily perceptible atmospheres (cf. Seamon 2020 & 2022). Seamon speaks of the "atmospheric qualities like sense of place" and says: "To clarify the lived subtlety of sense of place is one task of phenomenological research." (Seamon 2022). "Sense of place" could be paraphrased as "felt meaning of a place". At the same time, it also means “the sensuous ability to feel this meaning”. With reference to Relph (2009), Seamon distinguishes between the “genius loci” and the “sense of place” (in a narrower meaning). The former is described as “the singular qualities of a particular landscape or environment that infuse it with a unique ambience and character”; the latter as “the synaesthetic and largely unself-conscious facility of human beings to recognize, feel, and sense the uniqueness of a particular” place (Seamon 2022). “Genius loci” thus represents the intersubjectively perceived object side, “sense of place” the bodily felt subject side. If the layperson or phenomenologist hasn't developed and cultivated his sense of place, he isn't able to detect and analyse any genius loci; or, with Seamon's words: he will not be able to sense the “common presence”. The concept of “common presence” is similar to Schmitz's concept of a common impressive situations, but is less formal and closer to lived experience. Seamon is arguing that the relative togetherness of entities in space – material and human qualities as well – sustains an environmental “common presence” that emerges as a sensible quality shared by these local entities. And: „The common presence of a place refers to its degree of ‚life' and its environmental character“ (Seamon 2018, 87-90).

It is obvious that this “common presence” cannot be caught up theoretically, but only through phenomenological “field research”; this means: through “thick participation” (Spitteler 2014) and “thick description” (Geertz 1973). This point has been seen time and again, but seldom clearly enough (cf. Tilley 2004, 27f. see also: Rauh 2012 & 2018). Jürgen Hasse in particular has set new standards with his “micrologies of spatial experience”. He recorded his impressions in meticulous on-site protocols – and reflected on them neo-phenomenologically. “The great degree of differentiation” of his micrologies “owes itself exclusively to the logging explication of atmospheric experience in the actual situation of being-with.” (Hasse 2017, 170). In his micrologies, Hasse referred primarily to ordinary, everyday places – not genius loci sites. This should be reformed. Accordingly, his method of distanced participation would have to be deepened into thick participation (cf. chapter 9).

6.3 *My concept of anchored atmospheres and auratic places*

If I'm not mistaken, I was the first to make a clear distinction between place, landscape and geographical-climatic zone (in my diction: “earthscapes” cf. chapter 4.3) and to show that Christian Norberg-Schulz's genius loci types are landscape or even earthscape types and not place types. Be that as it may, Norberg-Schulz's phenomenology and typology is and remains a first and important (but incomplete) attempt (cf. Norberg-Schulz 1980; Kozljanič 2004b, 258, 309-320; Griffero 2020, 140f.). According to my concept places are delimitable, small-scale locations (usually smaller than a football field), landscapes are panoramatically overviewable, large-scale regions (that can extend to the horizon) and earthscapes are huge geographical units that reach much farther than the eye can see (unless you are in outer space). As I demonstrated, atmospheres are not only quasi-climatic, fragrance-like, and spatially flowing 'diffusivities'. In concretising Schmitz, I identified that atmospheres are also condensed in a place and anchored in specific things and shapes. They can be experienced empathically, perceived physiognomically and communicated intersubjectively as thingly phenomena within affective-expressive encounter-situations (see Kozljanič 2004b, 330-335; Kozljanič 2012, 82-87).

These thingly phenomena and their shape are of crucial importance: if they are removed or technically dominated or superstructured, the atmosphere and character of the place will seriously be damaged. This happened to the Loreley, which has almost completely lost its threatening and mysterious atmosphere due to construction works: Railway line and tunnel in 1862, then quay and road, blasting of the river stones in the 1930s, construction of an open-air Nazi theatre on the summit 1935-39. A modern equivalent of the Roman term "numen" seems to be "aura", coined by Walter Benjamin in his famous essay "The Work of Art in the Age of Mechanical Reproduction" (1935): Every original and ingenious work of art has or can have a special "aura" – and loses it through reproduction and instrumentalisation. As I showed above in Chapter 5, the same can be said of original and ingenious places (like the ruins of Judengasse in Frankfurt). I therefore distinguish between five key components: 1. The uniqueness of the object. 2. The sensuous presence of the original. 3. The aesthetic and contemplative distance of the viewer. 4. The singular-auratic emotional agitation of the beholder; or, in other words: the enchantment of the contemplator by a mesmerizing or "ecstatic" thing.[3] 5. The multi-layered testimony-character of a memorial-like object.

3 Cf. the term "ecstasy of things" in relation to the term "aura" in Böhme (1995, 25-34, 166-172); see also: Böhme (2006, 149-150).

6.4 *Tonino Griffero and his pathic-aesthetic concept of intense atmospheric affordances*

The most recent and forward-looking contribution to a phenomenology of the genius loci is by Tonino Griffero. He too revises and concretises Schmitz's concept of atmospheres. Central to this is the term "affordances", which he adopts from the US psychologist James J. Gibson: "The *affordances* of the environment are what it *offers* the animal, what it *provides* or *furnishes*, either for good or ill. […] I mean by it something that refers to both the environment and the animal in a way that no existing term does. It implies the complementarity of the animal and the environment." (Gibson 1979, 127; cf. Griffero 2020, 81). However, Griffero then places aesthetic-pathic suggestions alongside the pragmatic-active options for action. This rounds off the ethologistic concept of "affordances", deepens it and gives it enormous phenomenological relevance. "Affordances" now means: stimulating options and 'mesmerizing' significances radiating from the people-place-interactions, people-place-correlations, people-place-empathies. E. g. kinetic suggestions or psychic image-initiations, correspondence offers or resonance possibilities, bodily-activating triggers or bodily-pathic resonances. These affordances arise from the situations and their effects, whether they are embedded features, functional dispositions or expressive atmospheres. Either way, Griffero's focus is on "pathic aesthetics". He is primarily concerned with "pathic affordances, responsible for our spontaneous-intuitive evaluations".

His forward-looking idea is,

> that atmospheres function as (amodal) affordances, i. e. as ecological invitations or meanings that are ontologically rooted in things and quasi-things, namely as demands that are not only pragmatic-behavioural and visual. While the environment can invite a certain action or even urge a person to do something, to an atmospheric affordance indeed one does not necessarily react with a given behaviour. […] one may also react to atmospheric qualia with an (also aesthetic) distance, in the sense that we can feel the atmosphere of a chair, for example, both by sitting on it and by perceiving it from a distance. (Griffero 2020, 101).

Genius loci sites are thus marked by auratic and particularly "bubbly" (cf. Sloterdijk 2011) expressive-atmospheric affordances: "My hypothesis is therefore that a place has its own *genius* only if (when and where) it radiates an intense and authoritative specific atmosphere." (Griffero 2020, 145; cf. Tedeschini 2017).

7.
Problems of Approach and Methodology
Environmental (Neo)phenomenology, Thick Descriptions and Deep Hermeneutics

It is obvious that a monoperspectival approach cannot do justice to the phenomenon of the genius loci, neither lifeworldly nor methodologically. For example, I was already consciously working with a layer-historical method in 2004; creatively following Nicolai Hartmann's ontology of layers (or strata), which I, however, revised in terms of cultural history. These different historical layers correspond to specific modes of access (mythico-archaical, Olympic-mythical, Christian-medieval, modern-secular, rationalistic-technical, ...) and to specific human-place relationships (relationship of integration, relationship of structuring, relationship of superstructuring, relationship of excessive planning and overplanning – see Chapter 2). Griffero has taken up these approaches:

> When interacting with a place, even the same one, we can in fact a) abandon ourselves in an ecstatic-cultic way to it [in my words: mythico-archaical access based on a relationship of integration]; or b) intervene on it through material or ideal superstructures, as with the Christian reinterpretation and architectural refunctionalization of previously pagan sacred places [in my words: Christian-medieval or modern-secular access based on a relationship of structuring]; finally c) we can flatten it, making it perfectly suited to host new architectural projects (almost) totally unrelated to the character of the place [in my words: rationalistic-technical access based on a relationship of excessive planning]. Only in the first case do we have a full experience of the *genius loci* (Griffero 2020, 144).

But as I would add, the mythico-archaical approach is, nowadays, very often ideologically superstructured and civilisation-process-related overlayered. Usually we can't just jump in, we have to remove the upper layers first, get through them, deepen our experience. What we need is a kind of felt and mental archaeology of our being-in-the-world, of our being-in-place.

7.1 *Seamon's triadic model: environmental ensemble, people-in-place, common presence*

The person who has so far worked most stringently with a multi-perspectival procedure, both theoretically and practically, is David Seamon. He has shown that the genius loci can best be analysed and explicated in a triadic procedure. This procedure is based on pre-scientific lifeworld experiences and works with an interdisciplinary and multi-perspectival methodology. In a creative and phenomenological continuation of John G. Bennett's "Elementary Systematics" (Bennett 1993), he pointed out that at least three perspectives were necessary: one topographical, one sociographical and one atmospheric. As Seamon says: place experience might best be understood via the relationship between three intertwined components: environmental ensemble, people-in-place, and common presence: "The keyword is togetherness, whereby the environmental and human elements of place are together (or not) in a mode of belonging (or not) that supports (or undermines) the life and wholeness of the place." (Seamon 2018, 90). It was Tomáš Valena who uttered a second important keyword: relatedness: "Man has a primal need to relate the objects and phenomena of his world to each other." Through a reciprocal and multi-layered trilogue between man, building and place, "site-bound" relationships are created (Valena 2009, 148).

7.2 *Hasse and Kozljanič: Spontaneously lived, aesthetically experienced and memorised-sedimented space*

In 2010, Hasse and I have pointed to three important spatial modes of enquiry: Living, experiencing, remembering. In relation to lived space (= *living*), bodily experience is primarily expressive-atmospheric and rooted in (often) impressive-present situations. The sensuous-aesthetic *experience* is shaped – especially in relation to landscape perception – by temporal categories of life experience and by aesthetic forms of perception that have become historical. In addition to the impressive-present situations, the segmented-complex situations provide the decisive basis here. In mental-cultural *remembering*, the associative meanings and narrated stories sedimented in socio-cultural memory come to the fore. The spatial experience thus acquires new historical and/or biographical and immaterial (semantic) dimensions. It is through them that significant places of memory become accessible. These mnemotopes are less anchored in impressive situations and only partially in segmented situations (Hasse, Kozljanič 2010). I would therefore like to conceptually clarify an idea already expressed at that time: and speak of sedimented-complex situations. They can only be caught up through a cultural-historical-semantic layer analysis – which goes beyond the purely phenomenological approach. Just as the archaeologist digs deeper into the earth and uncovers layer after layer, the cultural scientist should dig deeper into the social and cultural (local) memory, layer by layer (Kozljanič 2010b).

In doing so, we should not forget: History is based on stories. These stories are – due to the still powerful and power-obsessed patriarchal tradition – mainly his-stories. Stories of the Western victors and bellicists. Stories of a predatory capitalism. Stories of hegemonic masculinity. We need to deconstruct them and supplement them with her-stories. And we should not forget that his-stories and her-stories have always storeys. This means that they are multi-layered stories.

Parallel to this diachronic layer analysis, however, the synchronic semantic relational structures – which emanate from or lead to the place – must also be examined. Layer analysis and relational analysis are equally important. One goes into depth, the other into breadth.

How these different approaches and perspectives can be integrated into an overarching theory without loss of differentiation is still unclear today. Seamon's interdisciplinary and multi-perspective methodology seems to me to be the most suitable. For it does not end with a triadic model, but is open and expandable upwards. With and after Seamon, a tedradic, pentadic or even hexadic extension would be conceivable and feasible.

7.3 *Christopher Tilley: „make writing a voice for the stones, the places and the landscapes"*

Christopher Tilley (1955-2024) has made some crucial points regarding the first two of the three important spatial modes of investigation mentioned in section 7.2, namely living and experiencing. Until 2022, Tilley had been Professor of Anthropology and Archaeology at University College London. He was not only a professor, he was also a courageous person. Because it takes courage to stand up against the mainstream of 'landscape scholasticism'. But that is exactly what Tilley did when he wrote:

> In virtually all the academic literature it is quite striking how disembodied written landscapes become. This is because virtually everything written about landscape

> is not only written on paper; it is principally derived from paper. Landscape is not bodily experienced [...] What we are left with is paper landscapes, paper perspectives, and Merleau-Ponty's own philosophical work is no exception. [...] However, what they all have in common is the lack of any embodied perspective: they can only provide us with abstract models for thinking landscapes rather than models *of* landscapes as they are sensuously lived. [...] In a text all we can hope to to do is to evoke the sensuous qualities of place and landscape in a multisensorial way through our choice and use of words and the types of narrative structure employed, and this is the task of a richly textured carnal phenomenological 'thick' description in which we truly attempt to reflect on the character of our experience, as opposed to a thin and sensorily impoverished 'analytical' account. Such account must, of necessity, exploit the tropic or metaphoric nature of language, avoiding a deadened and deadening literalism, to make writing a voice for the stones, the places and the landscapes in which we are bodily immersed. (Tilley 2004, 27f.)

But Tilley does not stop there. In a similarly radical manner to Klages, Schmitz, Griffero or myself, he attempts to connect with the mythico-archaical experience of place:

> What I have been suggesting is that rather than regarding things, places or landscapes primarily as systems of signs, or as texts or discourses which encode meaning and reflect social identities in various ways, we can regard them as *agents* which actively produce that identity. In other words we need to think about places and landscapes animistically, in an analogous manner to the way in which we like to think about persons, as entities who can and do make a difference. (Ibid. 31).

Tilley's conclusions are among the most concise written on this subject. These conclusions would be less valid if Tilley had not been inspired by David Abram and his profound book 'The Spell of the Sensuous' (Abram 1996).[1] It is therefore worth quoting them in detail:

> 1. A phenomenological approach to landscape and place, as discussed here, using the framework of Merleau-Ponty's thought and interpretations of this thought by others [most important: David Abram], is not a philosophical approach emphasizing the personal and the subjective. It is an approach emphasizing the intertwining of subject and object, things and persons, mind and body, places and Being in the world. The rejection of any possibility of an objective approach does not mean that we pass into a realm of personal subjectivity, because meaning is grounded in the sensuous embodied relation between persons and the world, an invariant ontological ground for all feeling

1 See my review of David Abram's book in: Philosophischer Literaturanzeiger (2012), 65, 3, 245-255.

and all knowing taking place through persons with similar bodies.

2. Any study begins with lived experience, being there, in the world. It must necessarily be embodied, centred in a body opening itself to the world, a carnal relationship. The exploitation of basic bodily dyads provides one entry point into the study of place and landscape. A concentric grades sense of place and landscape provides another basic way in which meaning may be explored. Both originate in the body and extend outwards.

3. Perceptual meanings of place and landscape are constituted as gestalts, themes against horizons, to which the human body and the external world both contribute, a lived structure of experience formed through engagement and interaction in which the body-subject and the world flow into each other and form a part of each other. […] There is a dialectical exchange between the embodied structures of the engaged perceiver and the structures of that which is perceived.

4. This involves a dehiscence, an opening of my body to things, a reversible relationship between touching and being touched, myself and other, the effect of myself on things and those things on me.

5. In an experiential relationship with things, there is always a chiasm, an intertwining between 'outside' and 'inside', which mediate each other but never totally fuse. So my body is in contact with the world but still separate from it. My body experiences from the inside but opens itself to the outside. Since, as an embodied observer, I perceive the world through a set of frameworks which are habitual and grounded in the body, to a certain extent anonymous, these frameworks cease to be mine alone and are not therefore 'personal'. They are, however, both objective and subjective insofar as they simultaneously stem from my own body. [...]

6. Our primordial experience is inherently animistic, disclosing a field of phenomena that are all potentially animate and expressive because our perception involves the reversibility born out of our participation in the world.

7. Direct prereflective perception is inherently synaesthetic, disclosing the things and elements that surround us not as inert objects but as expressive subjects of experience, born out of our multidimensional sensorial participation in the world.

8. There is a fundamental temporal dimension to the body, place and landscape carried through movement and sedimented into what places and landscapes are and how we experience them.

9. […] It is in the context of a needful body reaching out to the world that meaning and significance are found. The manner in which we experience place and landscape is, however, forever unfinished, uncertain and therefore ambiguous. The ambiguity inherent to both that which we investigate (place, landscape) and how we perceive is not a problem for analysis. Instead it provides an inexhaustible field of affordances for us.

10. The aim of a phenomenological analysis is to produce a fresh understanding of place and landscape through an evocative thick linguistic redescription […]. This involves attempting to exploit to the full the tropic nature of our language […]. The mode of expression must resonate with that which it seeks to express. (Tilley 2004, 29f.).

8.
Aesthetic Revision of Hermann Schmitz's Neophenomenological Method

8.1 *The decisive neophenomenological requirement: explication of unintended, spontaneous first-hand life experience*

I will now conclude by examining Hermann Schmitz's phenomenological method. My guiding question is: Does his method meet his own neo-phenomenological requirements? The most crucial requirement is to provide a phenomenological explication of spontaneous and, as far as possible, authentic life experiences: "i. e. what really happens to people in a felt manner without them having intentionally constructed it. Spontaneous life experience is the ultimate yardstick for the justification of assertions" (Schmitz 2019, 48). With the phrase "what really happens to people in a felt manner," Schmitz addresses another central point: what he calls "affective involvement." Ultimately, this is the most decisive guarantee that something really affects me, that something is effective and real in this sense. Without "affective involvement," there is no deeper sense of reality (cf. ibid. 62f.).

For Schmitz, phenomenology is "a confirmation of what we actually mean when we use meaningful expressions, what their 'place in life' is." (Schmitz 1980, 16). He is concerned with uncovering the "involuntary impressions" (Schmitz 1995, 33; i. e. especially 'impressive-present situations') underlying scientific "theories" and philosophical "ideas" in order to "examine the basis of experience more closely" (Schmitz 1967, 6). By gaining and examining this experiential basis, Schmitz aims to "break through the overly abstract basis of the dominant European intellectual culture [...] and approach involuntary life experience with comprehending sensitivity" (Schmitz 2003, 7). Phenomenology means: "Working in constant contact with unbiased life experience" (ibid. 21). According to Schmitz, we can only achieve this contact when we unreservedly engage with experiences in the bodily presence and in unadulterated, involuntary situations. The decisive factor here is our own body and our own bodily capacity for feeling, perception and communication – which is conveyed atmospherically, essentially, though not exclusively, through emotions and feelings.

> New Phenomenology takes upon itself the task of making philosophy able to make sense of involuntary life experiences. Involuntary life experiences are everything distinctly happening to human beings without specific planning. They constitute the one true base for the vindication of our claims. There is nothing else standing against the arbitrariness of constructions. This is why New Phenomenology defends every involuntary mode of experience that it can discover against systematic and pre-established reframing. For instance, this is what happens when feelings are considered as private states of the soul and therefore, as everything psychic (*seelisch*), as something spaceless (*raumlos*). Against this view, I advance the thesis that *feelings are spatially effused atmospheres and forces that grasp us felt-bodily* (Schmitz 2023, 45).

8.2 *The neophenomenological method of Hermann Schmitz*

Let us take a look at Schmitz's method. He has formulated it in various places. I am referring to a somewhat more mature and later presentation.

> First and foremost is the requirement to build exclusively on relatively trivial life experience (which is accessible to everyone at any time, either fresh in their minds or in their memories), to express this in precise terms, and to combine these terms only in a logically clear manner in order to reconstruct areas of experience that may themselves be trivial [...] or esoteric. (Schmitz 1995, 33).

This requires "phenomenological self-control". This self-control is ensured by the three-stage method.

> 1. Descriptive stage: A subject area is highlighted from relatively trivial life experience in the specified sense by describing it as accurately as possible using everyday language. 2. Analytical stage: Recurring and intertwining basic features of the area are identified and defined in terms of terminology. 3. Combinatorial stage: Through the appropriate combination of the basic features in the defined terminology, complex components of the area are reconstructed, also as a test of the adequacy of the analysis. (Ibid.)

Secondly, Schmitz continues, it is important to clarify the involuntary impressions of life experience in a content-based and 'factual' manner: "For phenomenology is the endeavour to expose the hard ground of phenomena by systematically peeling away all assumptions that depend on arbitrariness, namely the states of affairs that must be recognised as such because they cannot seriously be disputed." (Ibid. 34).

It is interesting to see that and how Schmitz here – contrary to his own claim to be as close as possible to life experiences, which are accessible above all in impressive-present situations in a state of affective involvement – methodically distances himself from these life experiences. How he abstracts from them and disregards them. I will demonstrate and justify this with three objections.

1. The exclusion of affective involvement from the phenomenological investigation process is particularly disconcerting. As Schmitz repeatedly demonstrated, access to involuntary life experiences is only possible through bodily-affective involvement. When something affects or touches me bodily-affectively or bodily-emotionally in everyday life, it is usually anything but trivial for me at that moment. Nor is it "accessible to everyone at any time, either fresh in their minds or in their memories". At that moment, it is only accessible to me or to the people (and living beings) *involved in shared situations*. But this insight, which is crucial to Schmitz's phenomenology, does not appear here. Neither in the three-stage method nor in the states-of-affairs-clarification.

2. Schmitz does not usually refer to his depictions of phenomena as descriptions, but rather as *explications*. In a (simplistic) description, a phenomenon is essentially 'read' and 'traced' with words. Schmitz wants to distance himself from this, which is why he uses the term explication. When he speaks of explication, he means, first, that he sensitively analyses the diffuse, holistic situation in which a phenomenon presents itself to him and, second, that he interprets it in a phenomenon-oriented manner: at a high and differentiated meta-linguistic level – but nevertheless close to the native language. As important as his phenomenological 'explication' is in other contexts, it does not appear here. Here, he refers only to a three-step method of description, analysis and combination. This is remarkable and disturbing at the same time.

3. Even more surprising is that, at the point where it would actually be necessary to "approach involuntary life experience with comprehending sensitivity" (Schmitz 2003, 7) – which can only mean: *through sensitive explication of situations* – Schmitz advocates "to expose the hard ground of phenomena by systematically peeling away all assumptions that depend on arbitrariness", in order to get to "the states of affairs" (Schmitz 1995, 34). The expression "hard ground of phenomena" reminds us not only of the (reductionist) 'hard facts' of natural sciences; the concept of "states of affairs" also reminds us of abstract and formalistic scientific ways of

speaking and thinking. I recall: Shouldn't phenomena be accessible to Schmitz only in internally diffuse, holistic situations? And then to those being linguistically and interpretatively extracted in a kind of cautious and appropriate "explication," but in such a way that the situations are not violently altered? Where is the "hard ground" in this? Is it appropriate to call these sensitively extracted explications "states of affairs"? They are anything but 'factual' "states of affairs".

As I already mentioned in Chapter 6.1: Schmitz shows that genuine life experiences not necessarily require or presuppose abstract superconcepts nor singled-out *hard facts* (Schmitz: "objective facts"), but firstly holistic and meaningful *situations*. According to Schmitz, these situations are the original objects of perception, the original phenomena (cf. Schmitz 2019, 73-75; and Großheim 2018). They appear as bodily felt *interwoven ambiance-ensembles* with internally diffuse but rich structure. Although they occur only in a subjective state of affective involvement, they are more than subjective: they are intersubjective *gentle facts* (Schmitz: "subjective facts") concerning reality. It is the task of the phenomenologist, to examine, distinguish and explicate these initially diffuse ensembles. According to Schmitz, poetry is a "gentle explication of situations" (Schmitz 1995, 461f.). The phenomenologist should also gently verbalise and explicate situations – but, as far as possible, systematically and unmetaphorically (cf. Schmitz 2019).

Although Schmitz only occasionally hints at this, it is clear that this process of explication is highly hermeneutic. Hermeneutics (the interpretation of meaning) aims to shed light on the rich but diffuse holistic situations in terms of their concrete meanings and semantic relationships. Here, too, the purely phenomenological approach reaches its limits. For any hermeneutic interpretation – if it is to deserve its name – cannot do without creativity. In and with language, the hermeneuticist traces the underlying situation – but in doing so, creatively reinterprets it. Just as a landscape painter perceives a landscape as a subjective fact and reproduces/reinvents it in a picture: depicting some things more clearly and in greater detail than can be experienced in nature (intersubjectively); but still in such a way that it can in principle be experienced in nature – albeit with a decisive element of creative personal interpretation.

But all that seems forgotten at the very moment when Schmitz talks about his 'pure' phenomenological method. Based on my objections, it would be worth seriously asking whether Schmitz succeeded, on a methodological level, in breaking "through the overly abstract basis of the dominant European intellectual culture [...] and approach involuntary life experience with comprehending sensitivity" (Schmitz 2003, 7). I would say: in many of his concrete phenomenological investigations, he has certainly succeeded in doing so, but not on a methodological level.

I ask: does it make sense at all – especially with regard to the genius loci theme, but possibly also in general – to search for a 'pure' phenomenological-formalistic method? I don't think so. As I already mentioned in Chapter 7: It is obvious that a monoperspectival approach cannot do justice to the phenomenon of the genius loci, neither lifeworldly nor methodologically. And, like David Seamon, we should work with a multi-perspective approach, both theoretically and practically.

Last but not least we should not forget what Tilley stated in his conclusions 9 and 10: that we are dealing with ambiguous situations in which phenomena arise that can only be translated to a very limited extent from the bodily-felt sphere into an unembodied and purely formalistic meta-language. And that it is not the task of phenomenological analysis to resolve these ambiguities by over-formalising the phenomena, but rather to explicate them hermeneutically, creatively and metaphorically:

> It is in the context of a needful body reaching out to the world that meaning and significance are found. The manner in which we experience place and landscape is, however, forever unfinished, uncertain and therefore ambiguous. The ambiguity inherent to both that which we investigate (place, landscape) and how we perceive is not a problem for analysis. Instead it provides an inexhaustible field of affordances for us. [...] The aim of a phenomenological analysis is to produce a fresh understanding of place and landscape through an evocative thick linguistic redescription [...]. This involves attempting to exploit to the full the tropic nature of our language [...]. The mode of expression must resonate with that which it seeks to express. (Tilley 2004, 30).

8.3 *The aesthetic and hermeneutic revision of the phenomenological method*

Ingenious and comprehensive as Schmitz is, he has already addressed and examined the problem I have developed and substantiated here in a different but very similar context – that of 'aesthetic devotion' (cf. Schmitz 1977, 641-645; Schmitz 1995, 479-484).

The problem is: How should phenomenologists relate to basic life experiences? And this in such a way that, firstly, they do not lose bodily and life-related contact, but also, secondly, gain sufficient distance to enable insight into and overview of involuntary life experiences and, thirdly, are able to maintain a (oscillating) balance between personal regression and personal emancipation that is conducive to sensitive explication (of the respective situations), perhaps a little closer to emancipation, but without in any way crossing over into abstractionist, present-suspending meta-distance.

With regard to Greek gods as moving and overwhelming atmospheric forces, Schmitz says: "In order to assert oneself in the concert of competing emotional forces without denying one's emotions – which would be dishonest and sterile – humans need a distance in their emotions that is granted to them in aesthetic devotion." (Schmitz 1998, 97). For Schmitz, this statement applies not only to ancient Greeks, but also to us today – it applies in general. (In another article, I showed the extent to which aesthetic devotion can still reach mythical depths today; Kozljanič 2019).

Aesthetic devotion has two sides: one that is more pathic and receptive (pathic aesthetic perception, or rather aesthetic contemplation) and one that is more active and interactive (aesthetic behaviour and creativity). Aesthetic devotion should not be confused with what Schmitz calls "aesthetic pleasure": arbitrary, mostly mannered, but in any case existentially non-committal playing with aesthetic stimuli, ideas, and topoi.

According to Schmitz, this is a "dangerous game: as soon as aesthetic pleasure denies aesthetic devotion, it loses the ground that gives it stability and degenerates into snacking on stimuli". The "mere connoisseur" exchanges unbiased "naive emotion [...] for the habitus of detached participation from a lofty vantage point" (Schmitz 1998, 104).

Aristotle defines virtue as a happy medium between two extremes. In a similar way, Schmitz talks about aesthetic devotion (which he sees as close to a virtuous attitude and, further down, calls "aesthetic behaviour"):

Aesthetic devotion places

> people in a happy middle ground between two fates threatened by derailment. One is complete surrender to the stirring powers of emotion [...]. The other fate is the drying up of life through the loss of affective involvement, so that nothing is important anymore, because reason without emotion cannot set goals and becomes entangled in petty whims and selfish brittleness (ibid. 101).

Or, I might add, loses itself in rationally detached reductions and trans-empirical abstractionisms.

Aesthetic devotion opens up our methodical approach to the phenomena and contents of life we are seeking: "This happy middle ground grants humans ample access to moving and exciting, delicately touching and powerfully shaking forces and gives them the chance to assert themselves" (ibid. 103) and, particularly important in relation to our question, also the chance for "participatory observation".

Schmitz thus provides us with decisive moments of our sought-after phenomenological-hermeneutical method. We only need to put them together, add a few details, and relate them to the above discussion of methodology.

Interjection: Are we permitted to do so? Can we simply transfer what Schmitz says in the field of aesthetic philosophy to our phenomenological-methodological question? I think we can and should. Firstly, Schmitz is clear that aesthetic contemplation and aesthetic behaviour are important universal human processes that by far exceed and undermine the aesthetic field in the narrower (art-historical) sense. Secondly – as Schmitz is well aware – the 'phenomenon' (phainómenon) in the pre-introjectionist and non-reductionist way of seeing and speaking of the (pre-philosophical) Greeks, which he so highly values, is closely linked to sensuous-aesthetic perception (aísthesis). For the Greeks, this perception (aísthesis) is a 'sensuous phenomenon', i. e. something that 'appears in sensuous perception' (cf. Kozljanič 2024a). It is – prior to anything 'artistic' – the basis of all actual perception. And thirdly, the discipline of philosophical aesthetics is not only a theory of beauty and art, but – since Alexander Gottlieb Baumgarten (1714-1762) and his work 'Aesthetica' (1750) – above all a theory of sensuous appearances and sensuous cognition. Which brings us right to our problem.

What Schmitz says about all people therefore also applies (or even particularly applies) to phenomenologists, who devote themselves to the study of sensuous and emotionally affecting phenomena of life:

> Why should and must humans behave aesthetically? This question cannot be adequately answered by referring to the free play of the powers of the mind, imagination and reason, to which Kant attributes aesthetic judgement; for such psychogymnastic exercises have no weight [or life content] that could stand up to the seriousness of morality or religion [or seriously pursued phenomenology of life content]. The right to aesthetic behaviour is based on the human need to assert oneself in personal emancipation before the authority of feelings, without withdrawing from this authority or denying it (Schmitz 1998, 105).

How should we revise Schmitz's previous phenomenological method? Here is a brief, preliminary outline:

1. Seek out the non-prefabricated situation in which the involuntary life experiences to be investigated are possible and the phenomena appearing in them become accessible. Participate as intensely as possible in this situation

("thick participation" – Spitteler 2014). This creates a shared (sought-out, but not contrived!) impressive and intrinsically dynamic situation. That would be my answer to the question: 'Can I willingly experience the involuntary?' Yes. By seeking out precisely such intrinsically dynamic situations, but not contriving them, instead resonating with them and interacting with them.

2. Aesthetic contemplation and bodily-sensuous interaction with the phenomena that arise in the self-dynamic shared situation; in such a way that the state of 'thick participation' is dimmed in a certain way, the involvement decreases a little, and 'participatory observation' becomes possible at an aesthetic but still sympathetic distance.

3. Notes (experience logs, experience records) should be taken even during participatory observation. These are early and preliminary sketches of phenomena, empathetic interpretations and initial, sensitive explications, which can capture the freshness of the experiences in a differentiated way precisely because they still participate in the shared situation.

4. Later, at the desk in the studio, these field studies are subjected to further clarifying and differentiating writing work; however, this is done in such a way that the situational emotional and present references continue to play a key role, which is why metaphorical expressions should by no means be eliminated, but, on the contrary, should be used preferentially. It is not a matter of technical or logical deconstruction, construction and/or reconstruction, but rather of providing a "thick description" (Geertz 1973) of the phenomena experienced, using means that are as close as possible to the mother tongue, literary-humanistic, so to speak, writerly means.

Although many phenomenologists do not like to hear this, it should be stated clearly: phenomenology, often invoked since Edmund Husserl (1859-1938) in the sense of "Philosophy as a Rigorous Science' (in my opinion a *contradictio in adiecto*) is transformed here into an open and hermeneutical-aesthetic project: into a *phenomenology of life experience*. Such a phenomenology – once disparagingly referred to by Schmitz as 'picture book phenomenology' (Schmitz 1980, 24) – is closer to the English-language tradition of 'Nature Writing' (cf. Kozljanič 2018) than it might seem.

Nature Writing is a description of nature using literary and belletristic means. Or to put it more simply: Nature Writing is novel writing with nature, i.e. the main characters are landscapes, places, flora and fauna – and the author amongst them. Nature Writers can be poets and novelists, but also natural and cultural scientists who have retained a 'literary' and 'natural' sensibility. For example, there are excellent literary descriptions of nature by Mark Twain, Adalbert Stifter, Jack London, John Muir and Gary Snyder (cf. Gilbar 1998); but also by Alexander von Humboldt („Ansichten der Natur") and Ernesto Grassi („Reisen ohne anzukommen. Südamerikanische Meditationen"). Or, to name a book that I personally like very much: Rachel Carson's „The Sense of Wonder. A Celebration of Nature for Parents and Children".

Nature Writing tends to focus on large, distinctive landscapes and wild natural formations, but it does not shy away from small, idyllic places either. Many factors play a role in this orientation, such as subjective perceptions and geographical descriptions, human moods and climatic atmospheres, acquired knowledge of nature and botanical conditions, literary talent and on-site experiences, personal and cultural worldviews, and natural spatial associations and meanings. The decisive factor here is always that good and appealing descriptions of nature can only be created when the subjective and the objective, the poetic and the prosaic, the felt and the thought, the remembered and the experienced, romanticism and naturalism, passion and (writing) craftsmanship are allowed to interact and interplay. When successful, this interplay and interaction leads to a deeper encounter with nature, to a dialogue with nature in which the linguistic and pre-linguistic, the conscious and the unconscious are allowed to have their say. Don Scheese summed it up quite well:

> Nature writers celebrate place, the locale that becomes the locus of their affections. [...] Nature writing [...] is typically about the confluence of a place and a writer, the physical and the metaphysical, where – in the solitude of retreat from modernity – imagination and fact, the word and the world, merge. (Scheese 1996, 134.)

Both Nature Writing and Nature Phenomenology use the same approach to nature: the pathic-aesthetic one. And both, as Griffero made clear, need affective involvement and emotional dialogue in order to perceive and explain the impressive and sedimented natural situations:

> A pathic aesthetics [...] supposes that there is a meaning already sedimented outside of us. Therefore, its objective is to theoretically outline the contours of a sphere animated by feelings that are effused in the pre-dimensional space (atmospheres) and so quasi-thingly aggressive as to take possession of us (maybe even only hypodermically). This sphere appears necessarily and structurally vague [...], but it is still essential to give our lives the 'colour' that, alone, makes it meaningful or at least tolerable: only feeling certifies our existence and the reality of the world – that is, the active one (*Wirklichkeit*) and not the thingly one (*Realität*). The aesthetic 'justification' of our lives,

therefore, does not lie in works of art, in which – albeit thanks to an affective-somatic medium – the po(i)etic product prevails over the pathic, but in the much salient (and possibly contagious) everyday feelings. (Griffero 2020, IXf.)

9.
Pathic Dialogue with Nature
Resonating, Corresponding, Poeticising. Pathic Aesthetics in Bettina von Arnim and Ludwig Klages

In this chapter, the strictly archetypal, purely receptive and archaic-regressive pathic aesthetics of Ludwig Klages (1872-1956) are compared and differentiated with the playfully nature-loving, romantically creative pathic aesthetics of Bettina von Arnim (1785-1859). In Bettina von Arnim's work, it becomes particularly clear that and how bodily-atmospheric resonance relationships form the mood basis in the pathic dialogue with nature, on which emotional-gestalt-like and spiritual-metaphorical correspondence relationships are built up and enable a deep and imaginative approach to nature. A relational approach to nature that has not yet been sufficiently seen and appreciated; but which, in my opinion, shows the decisive structural criteria of Nature Writing and a pathic-aesthetic phenomenology: *resonating – corresponding – poeticising*.

Interpreting and extrapolating a short text by Bettina von Arnim, I would like to show how genius loci-related Nature Writing (cf. Kozljanič 2018) and genius loci-oriented phenomenology can interact and learn from one another.

9.1 *Bettina's lime tree talk*

First, here is the relevant passage from Bettina von Arnim's epistolary novel "Clemens Brentano's Spring Wreath: Woven from Letters of Youth, as He Himself Requested in Writing" (1844). This epistolary novel is largely based on actual correspondence and personal experiences.

> I want to gaze at the clouds and at the moon, from which the day is just bidding farewell, and I want to gaze until I discover another world, and when I have found it, then no tear shall enviously darken the splendour in which my soul reflects its colours! –
>
> And what are you whispering in my ear, lime tree? – 'Green, green is the delicate colour of peace of mind, green in the evening light is the cradle of

dreams! And every branch cradles a dream, and my leaves rustle in the web of dreams, and it beckons you!' –

Oh, be silent, lime tree, it is night-time, the stars glitter through your foliage and speak of other things; and it trickles through my bones! – In future, intuition shall be my seer, and when I have lent her the tones of my loving sorrow to paint the swelling and sinking of her yearning power, she shall comfort me again, she who, an eternal sea, rolls all tears of melancholy in her waves until, cleansed of gloom, they rise like electric fire from the bosom of the waves. –

'Oh, you!' whispers the lime tree, 'don't be conceited, that won't break the spell.'

I'm not listening to you, lime tree, I'm listening to the stars up there! I hear music, their light melting into the dark night sky, their rays clinking together in dance.

'Whatever you want with your lofty feelings,' says the lime tree again; 'they don't reach up there, come under my crown, it will shake its dew on you, so that I feel you are blessed.'

Oh no, the stars sound louder and clearer, I hear them joyfully singing their harmonious kinship into the free air.

'Do not resist my whisper,' says the lime tree again, flattering me, and meaning, 'What is the music of the stars compared to this? – Think of me, you create my mind through your grasp of my nature, so that it winds itself around you again, as yours now winds itself around me, it shall touch you and always, until your soul learns to soar lightly and boldly to its own joy, in one breath of lovely speaking tones!'

What do you say, lime tree? – Is my understanding the playful soul of your mind? – Lime tree says: 'My soul trickles over to you with shivers because you like to think it. Thinking animates, all beings are coloured in the light of thought. What is the evening glow of your thoughts that they fly far across the field with it, and because you feel it. And if there were no thinking, no being would be animated anymore, and creation would sink silently into itself. Thinking animates, and every being resounds in its own playful colour in its light, whereby everything lives and believes itself to be immortal, and yet they depend only on the mind, which is thinking.

We believe ourselves to be alive, and yet the secret joy of becoming within us is because we resound in the mind that thinks us!' –

I say again: Think of me thus, lime tree, for I would not wish to mature in thought more beautifully than in the green shimmer of your leaves, kissed by the evening light, and I would not wish my mind to be carried away more nobly than in the scent of your blossoms!

The lime tree rustles in the wind and shakes itself, tickled by the kind words I have spoken to it, for this does not happen every day. (Arnim 1959, 37f.).

As can be seen from the letter to her brother Clemens Brentano, this dialogue with the lime tree (in all likelihood) stems from a real experience Bettina had. We do not know where in this letter the line between subjective experiences in real life and subsequent embellishments lies. But we do know that during her childhood and youth, Bettina had a very spontaneous and intense, loving and intimate relationship with other people, animals, plants and objects, with buildings, ruins, natural sites, rivers and landscapes (Kozljanič 2015a, 55f.). And that she managed to maintain this closeness to nature and spontaneity throughout her life. Throughout her life, Bettina had an intuitive and imaginative approach to herself as well as to the world with its countless beings, things and 'spirits' (cf. Lauterbach 1991, 20f.). In this context, Jutta Schlich aptly described Bettina's "mind-expanding [psychedelic] fantasy' (Schlich 2003, 67). Many people who knew Bettina and lived with her attested to this and were fascinated by her intuitive and imaginative, visionary nature – sometimes even overwhelmed by it.

In my opinion, interpreting and extrapolating Bettina's lime tree talk is worthwhile for at least three reasons: 1. Because of her aforementioned intuitive genius, which allows her to delve deeper and see further in many areas. 2. Because here we grasp, in a nutshell, so to speak, what I myself have discovered in practice, independently of Bettina, and call 'heart dialogue'. 3. Because in this talk, Bettina also addresses a crucial conflict: that between an archaic, regressive 'cosmic' pathicity (à la Ludwig Klages) and a more playful, locally and naturally connected, romantically creative pathicity (à la Bettina von Arnim).

To understand this conflict, it is necessary to take a look at Klages' concept of cosmic-ecstatic pathicity.

9.2 *Ludwig Klages' cosmic-ecstatic pathicity: absorbed by archetypal images*

As far as Klages' cosmic-ecstatic pathicity is concerned, there is good introductory and advanced literature (cf. Kozljanič 2004c, 149-186; Preußer 2015, 147-152; Griffero 2020, 3-20), which, within the framework of a 'pathic aesthetic,' opens up the narrowness and rigour of Klages' concept towards more playful and dialogical approaches.

Tonino Griffero summed it up well: Klages' concept has a clear tendency towards 'an almost vegetative pathicity of the soul and of rhythm' (Griffero 2020, VII). At the same time, Griffero distances himself – rightly, in my opinion – from the one-sided ecstatic pathos of the so-called Schwabing Cosmics, to whom Klages belonged in his younger years:

> the pathicity I refer to here does not necessarily imply the prophetic cosmicity that animated, with sometimes even caricatural outcomes, the bohème of Stefan George's circle in Munich, notoriously fascinated precisely by raptures of various nature (Ergriffenheit). More simply it is the acknowledgement that the aesthesiological receptivity that mainly characterizes our involuntary vital experiences and indirectly also the cognitive sphere – especially when it is involving to the point of 'cornering' the perceiver (who is therefore 'pre-occupied' with it) – still requires an adequate philosophical thematization. (Ibid. IX).

The following description and interpretation shows how Klages himself envisages the ecstatic-pathic approach to the reality of images:

> Let us imagine that somebody is contemplating a gleaming jewel (for example, in a shop window), and is held fast in the contemplation. Let there be present in him at first only the feeling of himself and the image of the jewel. And now let there happen what in full strength is granted only to few, although its rudiments, if the object be suitably chosen (for example, the sight of the setting sun or the form of the beloved) are, after all, known to everybody: the contemplating subject is 'absorbed' in the contemplated object. Then consciousness has become a mirror, empty except for the blaze of the jewel, and the feeling of Ego is extinguished before the supreme power of the image. A man who has become pure contemplation knows of no 'existence', and 'has forgotten' himself, and nevertheless is in a state of ardour compared with which the loftiest content of thought grows pale. He is freed from trammels of solitary existence and *becomes* the reflected content, and through it becomes that of which it is 'in itself' not part, but the vision and the symbol – the world, 'infinity', and universe. (Klages 1929, 218f.)

In this will-lessly devoted state of purely pathic and ecstatic contemplation – when the contemplating subject is absorbed by the contemplated object – access to the reality of the images is open. In Hermann Schmitz's words, the contemplating subject finds itself in a state of "personal regression" and "excorporation" that leads from narrowness to absolute vastness (cf. Schmitz 2023, 105-109). Klages speaks of a 'melting form of dissolving the self,' but also of 'dreamlike states': "We all know from memory that we lived as if in a dream: in the ‚self-forgetfulness' of a rare state of bliss, or when we were completely ‚lost' in the sight of the setting sun, utterly ‚absorbed' in the beckoning song of a flute, deeply ‚immersed' in the distant image of bygone days of youth." (Klages 2023, 43).

According to Klages, in such a dreamlike, archaic-regressive state, the ecstatic person becomes a visionary and sees images or archetypes. By these ecstatically seen 'images,' Klages means visionary archetypes of primordial reality that flash briefly and then never return in the same form (i.e., identical). In such unique moments, the viewer drinks, as Klages once metaphorically put it, "a full draught from the cup of the *Eros of distance*, which moves out of the tangible world of things into the intangible reality of images!" (Klages 2023, 59).

And although Klages repeatedly points out that this ecstatic, regressive image vision can be evoked by a wide variety of phenomena – he has already mentioned the shimmering jewel, a sunset, the figure of the beloved; elsewhere he mentions plants and animals, scents and sounds, spring storms, mountain ridges, ocean surf, sheet lightning ... (cf. Klages 2023, 32); – it is nevertheless clear to him that it is above all the great cosmic phenomena that ignite the cosmogonic-erotic vision. As he wrote in a letter to Franziska zu Reventlow in November 1901: "... The dark days of explosive showers, the great days of the flaming universe – they *have* blossomed for us, and how small and indifferent is human destiny when it does not burn in cosmic fires! …" (Klages 1944, 512).

That is why Klages can also say:

> Eros is called elemental or cosmic, insofar as the individual being seized by it experiences itself as pulsing and flooded by an electric current, as it where, which, similar to material magnetism and unconcerned about their barriers, makes the most distant souls feel each other in the connecting pull […]. And it is called cosmogonic, because it is a state of torrential fullness, according to which the internal – immediately also becoming manifest – instantly becomes an external, a world and an appearing reality. (Klages 2023, 33f.)

The enchanting view of the sparkling starry sky is therefore particularly important – paradigmatic, so to speak. For Klages, the starry sky is – in a wholly mythical sense – a symbol of the distant past, of the cosmic ground, of the rhythmically revolving universe, and also of the mythical ancestors:

> The Eros of past times is the Eros of distance. Because of this, Eros is ignited by incorporeal images and places the contemplating beholder in connection with the intangible world of stars, in such a way that the earthly life of those still present in light pulsate with the rhythm of the universe; therefore his bearers were literally 'cosmic' beings (Klages 2023, 94).

This star-like Eros of distance is expressed in condensed form in the poem 'Schöne Fremde' (A Beautiful Foreign Land) by Joseph von Eichendorff. This is why Klages quotes it (or rather the third stanza) at a crucial point: „All the stars shine on me / With love's glowing gaze. / The *distance* speaks drunkenly / As if destined for great happiness“ (Klages 2023, 60).[1]

9.3 *Basic principles of heart dialogue extrapolated from Bettina's lime tree talk*

Let us now take a look at Bettina's lime tree talk. It begins – very much in keeping with Klages' cosmic pathicity – with an ecstatic gaze at the evening sky and the stars. What is important in this first “cosmic” contact is an opening up, an affective-atmospheric resonance and, finally, the ecstatic enthrallment of the viewer. So, as a first step, I note:

9.3.1 *Cosmic resonance: Unconscious attunement, atmospheric resonance and ecstatic excorporation*

What dominates here, then, is a ‘pantheistic experience', a surrender to cosmic-oceanic atmospheres. As Schmitz puts it, it is – as when falling asleep or in some deep trance-like states – a matter of bodily expansion with a tendency towards “excorporation”. “Excorporation occurs when privative expansion drifts away [... and] follows suit into a formless, measureless vastness.” (Schmitz 2023, 103.) In a state of “personal regression”, the bodily self expands into cosmic-oceanic atmospheres and dissolves into them. „We can define excorporation as the felt-bodily communication engaging with such vastness.” (Ibid.)

It is precisely in this sense that Bettina is lured from the starry sky to excorporation. But the disembodiment does not quite succeed because the waving and whispering lime tree interferes. And Bettina initially finds this disturbing and enchantment-breaking. Bettina therefore rejects the lime tree: “Oh, be silent, lime tree, it is night-time, the stars glitter through your foliage and speak of other things; and it trickles through my bones!”

But the lime tree does not remain silent; it continues to persistently make demands, requests, “affordances” (Griffero 2020). The tree whispers urgently to Bettina: “‘Oh, you!' whispers the lime tree, ‘don't be conceited, that won't break the spell.'” Bettina replies:

---

1 In German: „Es funkeln auf mich alle Sterne / Mit glühendem Liebesblick. / Es redet trunken die *Ferne* / Wie von künftigem, großem Glück.“ (Klages 1963, 94; emphasis added by L. K.).

"I'm not listening to you, lime tree, I'm listening to the stars up there! I hear music, their light melting into the dark night sky, their rays clinking together in dance. 'Whatever you want with your lofty feelings,' says the lime tree again; 'they don't reach up there, come under my crown, it will shake its dew on you, so that I feel you are blessed.'" Bettina again: "Oh no, the stars sound louder and clearer, I hear them joyfully singing their harmonious kinship into the free air."

Just like Klages, Bettina wants to surrender to the "intuition" of "another world" when she looks at the night sky; this intuition "who, an eternal sea, rolls all tears of melancholy in her waves until, cleansed of gloom, they rise like electric fire from the bosom of the waves."

But Bettina knows more than just the "eros of distance" – she also knows and practises an "eros of tender closeness" (cf. Kozljanič 2025). And this is what Bettina's lime tree talk is about: the cosmogonic-erotic rapture of distance, the "high-flying" cosmic shivers are to be concretised and localised from the perspective of a human, emancipated (but nevertheless devoted and earth-loving) self. The 'pure' harmony of the spheres: it cannot suffice here. In place of a one-sided resonant absorption ("excorporation"), there should first be "reciprocal encorporation" and correspondence (stage 2); and then, building on this, a nature poetic dialogue (stage 3).

Jutta Schlich has also identified these three stages, albeit in a slightly different context (Bettina's reception of Hölderlin). Stage one would therefore be the stage of "symbiosis" (at which Hölderlin would remain), stage two that of (Bettina's) "presenting sympraxis", and stage three that of (Bettina's) "representing sympraxis".

> Sympraxis thus comes into its own in its difference from symbiosis, from exclusively passive identification with the object of attraction, as the specifically passive-active method of human and world knowledge. In relation to presenting sympraxis, representing sympraxis emphasises the active moment. (Schlich 2003, 82).

Purely atmospheric resonance may be good for setting the mood and getting in tune, but it should not lead to "excorporation" and "personal regression" (Schmitz). A finely tuned and oscillating balance between "personal regression" and "personal emancipation" is required. What is more, it requires a skilful balancing of aesthetic, devotional distance and physiognomic, situational closeness and tenderness: closeness to a concrete, expressive natural being (in Schmitz's words: "quasi-thing") in an "impressive situation". We need the 'greenness' of the lime tree, its 'foliage,' its earthy and local quality of pathic closeness. We should enter into a shared physiognomic situation with it. But how should we enter into a relationship and dialogue with a single natural being? How do we get from pathic distance to pathic closeness?

Bettina's Linde also gives a hint about this:

> Do not resist my whisper,' says the lime tree again, flattering me, and meaning, 'What is the music of the stars compared to this? – Think of me, you create my mind through your grasp of my nature, so that it winds itself around you again, as yours now winds itself around me, it shall touch you [...]!'

... But with this hint, the next (and next but one) stage already comes into view.

9.3.2 *Body-soul-correspondence: Preconscious encounter with a nature being in reciprocal encorporation (mediated by bridge-qualities)*

At this stage, Bettina slowly begins to escape the spell of cosmic ecstasy. The whispering lime tree helps her in this. It is not the steel-blue cold (untouchable) distance of the infinite night sky, but the green, sheltering, heart-warming closeness of the lime tree that is to take the lead. Instead of an enrapturing *Eros of distance*, a tenderly touching *Eros of closeness* now begins to dominate. It is now about a body-and-soul grasping, touching, feeling – and being touched. One's own animated body should expand slightly, up to the crown of the lime tree, into its foliage, but not up to the infinity of the starry sky.

Bettina intuitively practises here exactly what Schmitz referred to above as "aesthetic devotion": Aesthetic devotion places "people in a happy middle ground between two fates threatened by derailment. One is complete surrender to the stirring powers of emotion [...]. The other fate is the drying up of life through the loss of affective involvement" (Schmitz 1998, 101). The happy middle ground also exists between "personal regression" and "personal emancipation". As Bettina clearly expresses elsewhere: "I have grown to love beauty, I accepted it when it was given to me, only to give it away again immediately. Only in freedom, in existing for myself, do I enjoy life; and I will never seize anything for myself. I will give myself, but I will not be a prisoner." (Arnim 1959, 146).

Bettina's lime tree suggests a heart dialogue of allowing oneself to be touched, a dialogue of metaphorical heart thinking. In and through this dialogue, the lime tree is given a face. Its essence can now appear physiognomically and emotionally. As the lime tree already whispered: "Think of me, you create my spirit through your grasp of my nature, so that it winds itself around you again, as yours now winds itself around me, it shall touch you".

It is important to note at this point that the word 'thinking' means anything but a purely logical, mental mechanism of thinking and calculating. As mentioned above, it refers to thinking with the heart, a dialogue on the bodily and mental level of the heart. However, before this dialogue can take place (at stage 3), an encounter and exchange must have taken place at the bodily-soul level (stage 2). What Schmitz calls 'reciprocal encorporation' (cf. Schmitz 2023, 104) plays a key role in this.

> Encorporation happens when the felt body establishes a connection to a partner (external encorporation) or through an internal division (internal encorporation, pain, pregnancy) so that the vital drive now embraces both. It is not limited to felt bodies, but also occurs as the felt body establishes a similar connection with inanimate elements. This is made possible by bridge-qualities connected to the felt body, which can be perceived on one's own felt body and in the forms that it meets. These bridge-qualities are kinetic suggestions and synesthetic characters. Kinetic suggestions are anticipations or impressions of movements that are perceived and felt on our own felt body, produced by still or dynamic forms […]. Synesthetic characters are intermodal properties that often (through not always) bear the name of specific sense-qualities […]. They are for instance sharp, bright, soft, acute, light, hard, smooth, warm […]; they can describe the quality of colours, noises, perfumes […] (Schmitz 2023, 103).

9.3.3 *Metaphorical heart dialogue: Conscious dialogue with a natural being, through empathy and linguistic-metaphorical poeticising*

Bettina allows herself to be drawn in by what the lime tree and its leaves whisper, murmur and wave to her. Their "spirits" meet and entwine. An encounter takes place through reciprocal encorporation with the help of the bridge-qualities. When we now move from stage 2 to stage 3, something new is added, something that is more than bodily and belongs closer to the realm of soul and spirit: human language and human thinking through words, sentences and stories.

The key role played by bridge-qualities in stage 2 is now taken over in stage 3 by what I would like to call 'bridge-metaphors'. And in a similar way, I would like to say that thinking our way into the 'speechless' world of non-humans is made possible by bridge-metaphors that can spring from the heart dialogue between our own body-soul and the figures (gestalts) we encounter. Such bridge-metaphors translate the kinetic suggestions and synesthetic characters into human-poetic language. They lend a human voice to the non-verbal natural being (in our case, the lime tree).

How should we think about this lending-of-voices? The lime tree gives us another clue.

> Lime tree says: 'My soul trickles over to you with shivers because you like to think it. Thinking animates, all beings are coloured in the light of thought. What is the evening glow of your thoughts that they fly far across the field with it, and because you feel it. And if there were no thinking, no being would be animated anymore, and creation would sink silently into itself.

Here, too, it is clear that Bettina's lime tree does not mean rational thinking – separated from the heart – but rather heart-and-head thinking based on the body and soul, using bridge-metaphors. We could therefore translate: Lime tree says: 'My suggestions and synesthesia trickle over to you with shivers because you are able to encorporate them with your heart, empathise with them and translate them metaphorically. This leads to heart thinking, to heart dialogue: all beings are coloured by the light of heart thoughts. ... And if it were not for metaphorical heart thinking, no being would be animated anymore, and creation would sink silently into itself.'

I will not specifically justify here that this 'animating the lime tree' and 'lending it a voice' is anthropomorphisation but not anthropomorphism; that this 'thinking with the heart' is not a purely projective monologue, but an introjective-projective dialogue. A dialogue in which the human heart (as "felt-bodily isle"; cf. Kozljanič 2023c) remains in resonance with the lime tree, aligning its correspondence with the tree's bridge-qualities and affordances, its dialogue with the bridge-metaphors inspired by them. In and through this dialogue, the lime tree is given a voice. As a natural being, it can now also appear in human language. And this is equally important for Nature Writing and Phenomenology of Nature: both attempt to bring the world and its countless phenomena to life in a colourful and vibrant way using human language.

9.3.4 *"Another world" as a creative world of relationships and "the secret joy of becoming"*

If such a dialogue of the heart succeeds, then perhaps something can happen that Bettina only hints at here: we enter a poetic, dyadic space of the soul in which rational opposites become irrelevant – and the polar connection, the deep relationship between subject and object, between projection and introjection, between 'anthropomorphism' and 'naturalism', becomes apparent. According to Bettina, a shared spirit is formed, a heart-mind dyad of creative quality. In the words of the lime tree: "We believe ourselves to be alive, and yet the secret joy of becoming within us is because we resound in the mind that thinks us!"

This creative quality not only connects us to what Henri Bergson (1859-1941) called "creative evolution" ("l'évolution créatrice") and Anna-Teresa Tymieniecka (1923-2014) called 'unity-of-everything-there-is-alive,' (Tymieniecka 1990, 6; cf. Tymieniecka 1988) but also to the dynamics of becoming, the source of all vitality within and outside of us: the joy of life and becoming. By tracing and translating this dyad with poetic words, we also bear witness: a concrete, language-creative and world-illuminating testimony to a profound encounter and relationship between humans and nature.

At the beginning of their lime tree talk, Bettina said: "I want to gaze at the clouds and at the moon, from which the day is just bidding farewell, and I want to gaze until I discover another world, and when I have found it, then no tear shall enviously darken the splendour in which my soul reflects its colours!"

Interestingly, Bettina only discovered "another world" when she listened to the whispering of the lime tree and followed it. And it became clear that this other world is not a different, foreign, cosmic, infinitely distant world at all, but a world of its own, a real world, a world that is part of our own, a world within our world: the world that opens up to us when we allow our hearts to think.

The difference that Jutta Schlich identified between Bettina's poetic "sympraxis" and Hölderlin's poetic "symbiosis" can also be asserted – under more philosophical auspices – between Bettina and Klages: "In contrast to Hölderlin's authoritatively defined poetic legislative and judicial powers, Bettina's sympractical, passive-active, perceiving-judging poetic executive is, in a unique sense, *in statu nascendi.*" (Schlich 2003, 87).

Finally, here is Bettina again:

> I say again: Think of me thus, lime tree, for I would not wish to mature in thought more beautifully than in the green shimmer of your leaves, kissed by the evening light, and I would not wish my mind to be carried away more nobly than in the scent of your blossoms! The lime tree rustles in the wind and shakes itself, tickled by the kind words I have spoken to it, for this does not happen every day.

Picture credits

All photos are mine, R. J. K. Some of the historical images and graphics are in the public domain, while others were taken from my two-volume monograph “Der Geist eines Ortes” (Munich 2004). With kind permission from Albunea Verlag.

Bibliography

Abram, D. (1996). *The Spell of the Sensuous*. New York: Vintage.

Arnim, B. v. (1959). *Clemens Brentanos Frühlingskranz. Aus Jugendbriefen ihm geflochten, wie er selbst schriftlich verlangte* (1844). In G. Konrad (ed.). *Bettina von Arnim. Werke und Briefe,* vol 1. Darmstadt: WBG, 11-214.

Ash, H. B. (ed.). (1934). *Marcus Porcius Cato, "On Agriculture". Marcus Terentius Varro, "On Agriculture"*, trans. by W. D. Hooper. London: Heinemann.

Bachelard, G. (1964). *The Poetics of Space*. New York: Orion.

Bartetzko, D. (1992). *Damnatio memoriae – Der Börneplatz als Ort des kollektiven Vergessens*. In G. Heuberger (ed). *Stationen des Vergessens – Der Börneplatzkonflikt. Publikation zur Eröffnungsausstellung des Museums Judengasse*. Frankfurt a. M.: Jüdisches Museum.

Baumeister, T. (1991). *Höhenkult*. In E. Dassmann et al. (eds.), *Reallexikon für Antike und Christentum*, vol. 15. Stuttgart: Hiersemann, cols. 992-1015.

Benjamin, W. (1961). *Kurze Schatten*. In S. Unseld (ed.). *Walter Benjamin – Illuminationen. Ausgewählte Schriften*. Frankfurt a. M.: Suhrkamp, 319-328.

Benjamin, W. (1969). *The Work of Art in the Age of Mechanical Reproduction* (first version 1935). In H. Arendt (ed.). *Walter Benjamin – Illuminations*. New York: Schocken, 217-251. I am quoting from the MIT PDF (https://web.mit.edu/allanmc/www/benjamin.pdf), which is identical to the 1969 text but has its own page numbering.

Benjamin, W. (1980). *Das Kunstwerk im Zeitalter seiner technischen Reproduzierbarkeit* (third version 1939). In R. Tiedemann, H. Schweppenhäuser (eds.). *Walter Benjamin – Gesammelte Schriften,* vol 1, 2. Frankfurt a. M.: Suhrkamp, 471-508.

Benjamin, W. (1982). *Das Passagen-Werk* (1928-1929, 1934-1940). In R. Tiedemann, H. Schweppenhäuser (eds.). *Walter Benjamin – Gesammelte Schriften,* vol 5, 1. Frankfurt a. M.: Suhrkamp.

Bennett, J. G. (1993). *Elementary Systematics*. Ed. by D. Seamon. Santa Fe: Bennett Books.

Best, M. (ed.). (1988). *Der Frankfurter Börneplatz. Zur Archäologie eines politischen Konflikts*. Frankfurt a. M.: Fischer.

Birt, T. (1886-1890). *Genius*. In W. H. Roscher et al. (eds), *Ausführliches Lexikon der griechischen und römischen Mythologie*, 6 vols, here vol. 1.2 (Vorläufige Nachträge zu Bd. 1). Leipzig: Teubner, cols 1613-1625.

Bishop, P. (2018). *Ludwig Klages and the Philosophy of Life*. London-New York: Routledge.

Böhme, G. (1985). *Anthropologie in pragmatischer Hinsicht*. Frankfurt a. M.: Suhrkamp.

Böhme, G. (1989). *Für eine ökologische Naturästhetik*. Frankfurt a. M.: Suhrkamp.

Böhme, G. (1995). *Atmosphäre. Essays zur neuen Ästhetik*. Frankfurt a. M.: Suhrkamp.

Böhme, G. (1997). *Die Phänomenologie von Hermann Schmitz als Phänomenologie der Natur?* In G. Böhme u. G. Schiemann (eds.). *Phänomenologie der Natur*. Frankfurt a. M.: Suhrkamp, 133-148.

Böhme, G. (2006). *Architektur und Atmosphäre*. München: Fink.

Böhme, G. (2017a). *Atmospheric Architectures. The Aesthetics of Felt Spaces*. London-New York: Bloomsbury.

Böhme, G. (2017b). *Critique of Aesthetic Capitalism* (Atmospheric Spaces 1) Milan: Mimesis.

Bollnow, O. F. (1963). *Mensch und Raum*. Stuttgart: Kohlhammer.

Bollnow, O. F. (1980). *Die Welt der Symbole*. In G. Stephenson (ed.). *Leben und Tod in den Religionen – Symbol und Wirklichkeit*. Darmstadt: WBG.

Bollnow, O. F. (2020). *Human Space* (1963) (Atmospheric Spaces 8). Milan: Mimesis.

Borch, C. (ed.) (2014). Architectural Atmospheres. On the Experience and Politics of Architecture. Basel: Birkhäuser.

Buderath, B, Makowski, H. (1986). *Die Natur dem Menschen untertan. Ökologie im Spiegel der Landschaftsmalerei*. München: dtv.

Burckhardt, J. (1954). *Die Zeit Constantins des Großen* (1852/53). Frankfurt a. M.: Fischer.

Cassirer, E. (1944). *An Essay on Man. An introduction to a philosophy of human culture*. New Haven-London: Yale University Press, Milford.

Cassirer, E. (1960). *Was ist der Mensch? Versuch einer Philosophie der menschlichen Kultur*. Stuttgart: Kohlhammer.

Cassirer, E. (1994a). *Wesen und Wirkung des Symbolbegriffs*. Darmstadt: WBG.

Cassirer, E. (1994b). *Philosophie der symbolischen Formen*, vol. 1: Die Sprache (1923). Darmstadt: WBG.

Coates, G. J., Seamon, D. (1993). *Promoting a Foundational Ecology Practically Through Christopher Alexander's Pattern Language. The Example of Meadowcreek*. In D. Seamon (ed.). *Dwelling, Seeing, and Designing. Toward a Phenomenological Ecology*. New York: Suny, 331-354.

Curtius, E. R. (2013). *European Literature and the Latin Middle Ages* (1948). Princeton, Princeton University Press.

David, B., Wilson, M. (eds.). (2002). *Inscribed Landscapes. Marking and Making Place*. Honolulu: University of Hawai'i Press.

Demski, E. (1988): *Verraten und verbaut, Die Geschichte von der endgültigen Zerstörung der Frankfurter Judengasse*. In M. Best (ed.). *Der Frankfurter Börneplatz. Zur Archäologie eines politischen Konflikts*. Frankfurt a. M.: Fischer.

Duerr, H. P. (1985a) *Traumzeit. Über die Grenzen zwischen Wildnis und Zivilisation* (1979). Frankfurt a. M.: Suhrkamp.

Duerr, H. P. (1985b). *Dreamtime. Concerning the Boundary Between Wilderness and Civilization*. Oxford: Basil Blackwell.

Dürckheim, K. v. (2005). *Untersuchungen zum gelebten Raum* (1932). J. Hasse (ed.); with introductions by J. Hasse, A. Janson, H. Schmitz and K. Schultheiß. Frankfurt a. M.: Institut für Didaktik der Geographie.

Eck, D. v., Pijpers, R. (2016). „Encounters in Place Ballet. A Phenomenological Perspective on Older People's Walking Routines in an Urban Park". *Area*, 4. September 2016, 1-8.

Eisenhut, W. (1979a). *Cloacina*. In K. Ziegler, W. Sontheimer (eds.). *Der kleine Pauly. Lexikon der Antike*, vol 1. München: dtv, col. 1226.

Eisenhut, W. (1979b). *Iuturna*, In K. Ziegler, W. Sontheimer (eds.). *Der kleine Pauly. Lexikon der Antike*, vol 3. München: dtv, col. 25f.

Eliade, M. (1957). *Das Heilige und das Profane*. Hamburg: Rowohlt.

Eliade, M. (1961). *The Sacred and the Profane*. New York: Harper.

Feyerabend, P (1984). *Wissenschaft als Kunst*. Frankfurt a. M.: Suhrkamp.

Findeisen, H., Gehrts, H. (1983). *Die Schamanen*. München: Diederichs.

Freyer, H. (1990): *Landschaft und Geschichte* (1966). In G. Gröning, U. Herlyn (eds.). *Landschaftswahrnehmung und Landschaftserfahrung, Texte zur Konstitution und Rezeption von Natur als Landschaft*. München: Minerva, 44-47.

Gaßenhuber, R., Kozljanič, R. J. (2024). *In Search of Lost Liveliness. Nature, Dominance, Sense of Heart – The Long Way to Friendship with Nature*. (Atmospheric Spaces 15). Milan: Mimesis.

Garnham, H. L. (1985). *Maintaining the Spirit of Place. A Process for the Preservation of Town Character*, Mesa: PDA Publishers.

Geertz, C. (1973). *Thick Description: Toward an Interpretive Theory of Culture*. In *The Interpretation of Cultures: Selected Essays*, New York: Basic Books, 3-30.

Geisau, H. v. (1979): *Heroenkult*. In K. Ziegler, W. Sontheimer (eds.). *Der kleine Pauly. Lexikon der Antike*, vol 2. München: dtv, cols. 1103-1105.

Gibson, J. J. (1979). *The Theory of Affordances. The Ecological Approach to Visual Perception*. Boston: Houghton Mifflin Harcourt.

Gilbar, S. (ed.). (1998). *Natural State. A Literary Anthology of California Nature Writing*. Berkeley-Los Angeles-London: University of California Press.

Gimbutas, M. (1991). *The Civilization of the Goddess*. New York: Harper.

Grassi, E. (1940). *Der Beginn des modernen Denkens*. In E. Grassi (ed.). *Jahrbuch der geistigen Überlieferung*, vol. 1. Berlin: Küpper, 36-84.

Grätzel, S. (1997). *Das Verstummen der Natur*. Würzburg: Königshausen & Neumann.

Griffero, T. (2017). *Quasi-Things. The Paradigm of Atmospheres*. Albany: Suny.

Griffero, T., Moretti, G. (eds.). (2018). *Atmosphere/Atmospheres. Testing a New Paradigm* (Atmospheric Spaces 3). Milan: Mimesis.

Griffero, T. (2019). *Introduction. How Do You Find Yourself in Your Environment? Hermann Schmitz's New Phenomenology*. In Schmitz, H. *New Phenomenology. A Brief Introduction* (Atmospheric Spaces 6). Milan, Mimesis, 9-41.

Griffero, T. (2020). *Places, Affordances, Atmospheres. A Pathic Aesthetics.* London-New York: Routledge.

Griffero, T. (2021). *The Atmospheric "We". Moods and Collective Feelings* (Atmospheric Spaces 10). Milan: Mimesis.

Griffero, T. (2024). *Being a Lived Body. From a Neo-phenomenological Point of View*. London-New York: Routledge.

Groß, R. (1979). *Forum Romanum*. In K. Ziegler, W. Sontheimer (eds.). *Der kleine Pauly. Lexikon der Antike*, vol. 2. München: dtv, cols. 604-606.

Großheim, M. (2018). „Zu den Situationen selbst! Ein Vorschlag zur Reform der Phänomenologie" (To the Situations Themselves! Proposal for the Reformation of Phenomenology). *Synthesis Philosophica* 2, 303-325.

Gummere, R. M. (ed.). (1925). *Seneca. Ad Lucilium Epistulae Morales,* vol 1. London: Heinemann, 273-275.

Hanslik, R. (1979). *Nervae forum*. In K. Ziegler, W. Sontheimer (eds.). *Der kleine Pauly. Lexikon der Antike*, vol. 4. München: dtv, col. 76.

Hartmann, N. (1942). *Neue Wege der Ontologie*. Stuttgart: Kohlhammer.

Hartmann, N. (1953). *New Ways of Ontology*. Chicago: Regnery.

Hartmann, N. (1954). *Einführung in die Philosophie* (1949). Osnabrück: Hanckel.

Hasse, J. (2006). *Stadt als Heimat*. In H. Paetzold (ed.). *Integrale Stadtkultur*. Weimar: Verlag der Bauhaus-Universität, 156-179.

Hasse, J., Kozljanič, R. J. (2010). „Einleitung" (Gelebter, erfahrener und erinnerter Raum), *Jahrbuch für Lebensphilosophie*, 5, 9-21.

Hasse, J. (2017). *Die Aura des Einfachen. Mikrologien des räumlichen Erlebens*, vol. 1, Freiburg-München: Alber.

Hasse, J. (ed.). (2018). *Das Eigene und das Fremde. Heimat in Zeiten der Mobilität*. Freiburg-München: Alber.

Hauskeller, M. (1999). *Atmosphären in Natur und Kunst*. In M. Fechter, B. Wagner (eds.). *Gesellschaftliche Perspektiven. Zeit – Demokratie – Natur* (Jahrbuch der Hessischen Gesellschaft für Demokratie und Ökologie). Essen: Klartext, 97-108.

Heidegger, M. (1966). *Discourse on Thinking*. A Translation of "Gelassenheit" by J. M. Anderson, E. H. Freud, New York: Harper & Row. – German first edition: Heidegger, M. (1959). Gelassenheit. Pfullingen: Neske.

Heidegger, M. (2000a). *Bauen Wohnen Denken* (1951). In F.-W. v. Herrmann (ed.). *Gesamtausgabe I. Abteilung. Veröffentlichte Schriften 1910-1976. Band 7: Vorträge und Aufsätze*. Frankfurt a. M.: Klostermann, 145-164. – English Translation of „Building Dwelling Thinking" in: Heidegger, M. (1971). Poetry, Language, Thought. Trans. by A. Hofstadter. New York: Harper & Row, 143-161.

Heidegger, M. (2000b). *Aufenthalte* (1962). In C. Ochwadt (ed.). *Gesamtausgabe III. Abteilung. Unveröffentlichte Abhandlungen – Vorträge – Gedachtes. Bd. 75: Zu Hölderlin – Griechenlandreisen*. Frankfurt a. M.: Klostermann, 213-245.

Heuberger, G. (1988). *Schadensbegrenzung. Ein vorläufiges Resümee zur Rolle des Jüdischen Museums im Konflikt um den Börneplatz*. In M. Best (ed.). *Der Frankfurter Börneplatz. Zur Archäologie eines politischen Konflikts*. Frankfurt a. M.: Fischer.

Heuberger, G. (ed). (1992a). *Stationen des Vergessens – Der Börneplatzkonflikt. Publikation zur Eröffnungsausstellung des Museums Judengasse*. Frankfurt a. M.: Jüdisches Museum.

Heuberger, G. (ed). (1992b). *Museum Judengasse. Katalog zur Dauerausstellung des Jüdischen Museum im Auftrag des Magistrats der Stadt Frankfurt*, Frankfurt a. M.: Jüdisches Museum.

Hunt, J. D., Willis, P. (eds.). (1975). *The Genius of the Place. The English Landscape Garden 1620-1820*. London: Paul Elek.

Hunt, J. D. (2022). *Genius Loci. An Essay on the Meanings of Place*, London: Reaktion.

Ingold, T. (2020). *Hunting and Gathering as Ways of Perceiving the Environment*. In R. Ellen, K. Fukui (eds.). *Redefining Nature*. London-New York: Routledge, 117-156.

Joisten, K. (2012). *Heimat*. In S. Günzel, F. Kümmerling (eds.), *Lexikon der Raumphilosophie*. Darmstadt: WBG, 170-171.

Jones, W. H. S., Ormerod, H. A. (eds.). (1918). *Pausanias. Description of Greece*, 6. vols. London: Heinemann.

Jung, C. G. (1969a). *The Structure of the Psyche (Die Struktur der Seele)*. In H. Read et al. (eds.), *C. G. Jung. Collected Works*, vol. 8. Princeton: Princeton University Press, 139-158.

Jung, C. G. (1969b). *On the Nature of the Psyche* (*Theoretische Überlegungen zum Wesen des Psychischen*). In H. Read et al. (eds.), *C. G. Jung. Collected Works*, vol. 8. Princeton: Princeton University Press, 159-234.

Jung, C. G. (1969c). *Archetypes of the Collective Unconscious (Über die Archetypen des kollektiven Unbewussten)*. In H. Read et al. (eds.), C. G. Jung. Collected Works, vol. 9, 1. Princeton: Princeton University Press, 3-41.

Jung, C. G. (1970). Archaic Man (*Der archaische Mensch*). In H. Read et al. (eds.), *C. G. Jung. Collected Works*, vol. 10. Princeton: Princeton University Press, 50-73.

Klages, L. (1929). *The Science of Character* ("Die Grundlagen der Charakterkunde". Fifth & sixth ed; transl. by W. H. Johnston). London: Allen & Unwin.

Klages, L. (1936). *Grundlegung der Wissenschaft vom Ausdruck.* (Fifth, completely revised edition of „Ausdrucksbewegung und Gestaltungskraft", 1913). Leipzig: Barth.

Klages, L. (1944). *Rhythmen und Runen*. Leipzig: Barth.

Klages, L. (1963). *Vom kosmogonischen Eros* (1922). Bonn: Bouvier.

Klages, L. (1969). *Der Geist als Widersacher der Seele* (1929-32). In E. Frauchiger et al. (eds.). *Ludwig Klages – Sämtliche Werke,* vols. 1 & 2. Bonn: Bouvier.

Klages, L. (1974). *Vom kosmogonischen Eros* (1922). In E. Frauchiger et al. (eds.). *Ludwig Klages – Sämtliche Werke,* vol. 3. Bonn: Bouvier, 353-497.

Klages, L. (2023). *On Cosmogonic Eros.* ("Vom kosmogonischen Eros", 1922; transl. by K. J. Elliott). [No location specified] Anarch Books.

Korn, S. (1988). *Die Moral des Ortes.* In M. Best (ed.). *Der Frankfurter Börneplatz. Zur Archäologie eines politischen Konflikts*. Frankfurt a. M.: Fischer.

Kozljanič, R. J. (2001). *Kunst und Mythos. Lebensphilosophische Untersuchungen zu Ernesto Grassis Begriff der Urwirklichkeit.* Oldenburg: Igel.

Kozljanič, R. J. (2003). *Ernesto Grassi – Leben und Denken.* München: Fink.

Kozljanič, R. J. (2004a). *Der Geist eines Ortes: Kulturgeschichte und Phänomenologie des Genius Loci. Vol. 1: Antike – Mittelalter.* München: Albunea.

Kozljanič, R. J. (2004b). *Der Geist eines Ortes: Kulturgeschichte und Phänomenologie des Genius Loci. Vol. 2: Neuzeit – Gegenwart.* München: Albunea.

Kozljanič, R. J. (2004c). *Lebensphilosophie. Eine Einführung.* Stuttgart: Kohlhammer.

Kozljanič, R. J., Hasse, J. (2010a). „Einleitung“ (Gelebter, erfahrener und erinnerter Raum), *Jahrbuch für Lebensphilosophie*, 5, 9-21.

Kozljanič, R. J. (2010b). „Landschaft als physiognomisch-atmosphärisches und geistig-kulturelles Phänomen“. *Jahrbuch für Lebensphilosophie*, 5, 157-176.

Kozljanič, R. J. (2012). *Genius Loci and the Numen of a Place. A Mytho-Phenomenological Approach to the Archaic.* In P. Bishop (ed.). *The Archaic – The Past in the Present.* London-New York: Routledge, 69-92.

Kozljanič, R. J. (2015a): „Lebenskunst, Herzensweisheit und Liebesdenken bei Bettina von Arnim. Oder: Sein ungelebtes Leben tanzen“. *Jahrbuch für Lebensphilosophie*, 7, 55-74

Kozljanič, R. J. (2015b). *Leben, Wohnen, Fühlen. Oder: Von der beheimatenden Funktion ‚herzerwärmender' Orte.* In M. Großheim, A. K. Hild, C. Lagemann, N. Trčka (eds.). *Leib, Ort, Gefühl. Perspektiven der räumlichen Erfahrung.* Freiburg-München: Alber, 369-392.

Kozljanič, R. J. (2018). *“Nature Writing und Naturgefühl – Alexander von Humboldt”. Erleben & Lernen*, 2, 24-26.

Kozljanič, R. J. (2019). *Die heutige „Vision Quest“ als naturästhetische Kontemplation und säkularisierte religiöse Naturerfahrung.* In C. E. Herrera Krebber (ed.). *Ganzheitsmedizin IV: Der Weg zu einem neuen Bewusstsein der Heilung. (Holistic Medicine IV: The Way to a New Consciousness of Healing)*. Norderstedt: BoD, 245-274.

Kozljanič, R. J. (2022a). *Philanthropie als Maske der Macht. Corona, Gates und das Geschäft mit der Angst. Ein Handbuch zur Enttarnung antidemokratischer Netzwerke.* München: Albunea.

Kozljanič, R. J. (2022b). „Leben, lieben, aufbegehren – Sozialismus des befreiten Herzens“. *Jahrbuch für Lebensphilosophie*, 10, 11-100.

Kozljanič, R. J. (2023a). “Genius Loci”. *International Lexicon of Aesthetics.* https://lexicon.mimesisjournals.com/international_lexicon_of_aesthetics_item_detail.php?item_id=133 (accessed July 5, 2025).

Kozljanič, R. J. (2023b). “Toward a Phenomenology of Nature-made and Human-made Genius Loci”. *Environmental and Architectural Phenomenology*, 2, 17-24.

Kozljanič, R. J. (2023c). *Spüren, Berühren, Berührtwerden – Über die Schlüsselrolle der Herzgefühle bei leibseelischen Berührungen.* In C. Uzarewicz, R. Gugutzer, M. Uzarewicz, T. Latka (eds.). *Berühren und berührt werden. Zur Phänomenologie der Nähe,* Baden-Baden: Alber, 75-103.

Kozljanič, R. J. (2024a). *Die Befreiung der Phänomene. Über Schmitz und Klages zu einer (er)lebensphilosophischen Phänomenologie.* In M. Großheim, D.

Smiljanić (eds.). *Ludwig Klages und die Neue Phänomenologie*. Baden-Baden: Alber, 135-176.

Kozljanič, R. J., Gaßenhuber, R. (2024b). *In Search of Lost Liveliness. Nature, Dominance, Sense of Heart – The Long Way to Friendship with Nature.* (Atmospheric Spaces 15). Milan: Mimesis.

Kozljanič, R. J., Stöcker, H. (2024c). *Lebens-Liebes-Kunst und die Befreiung der Geschlechter. Texte zur solidarischen Emanzipation von Frauen und Männern.* München: Albunea.

Kozljanič, R. J. (2025). *Eros, Sexus, Lebenskunst. Liebeskonzepte bei Ludwig Klages und Franziska zu Reventlow.* In P. Bishop, P., H.-P. Preußer (eds.). *Eros und Sexus in der Philosophie von Ludwig Klages*. Würzburg: Königshausen & Neumann, 83-119.

Kriss, R., Kriss-Heinrich, H. (1955). *Peregrinatio Neohellenika., Wallfahrtswanderungen im heutigen Griechenland und in Unteritalien*, Wien: Verlag des Österreichischen Museums für Volkskunde.

Kufner, L. (1992). *Getaufte Götter. Heilige zwischen Mythos und Legende*. München: Pfeiffer.

Kuhlmann, D. (1998). *Der Geist des (W)ortes*. In https://www.cloud-cuckoo.net/openarchive/wolke/deu/Themen/982/Kuhlmann/kuhlmann_t.html (accessed June 29, 2025).

Kurz, O. (1953). "Huius Nympha Loci: A Pseudo-Classical Inscription and a Drawing by Dürer". *Journal of the Warburg and Courtauld Institutes*, 16, 3/4, 171-177.

Lauterbach, H. (1989). *Universalistisches Naturrecht und Religionsphilosophie in Bettine v. Arnims Schrift ‚Die Klosterbeere. Zum Andenken an die Frankfurter Judengasse' aus den ‚Gesprächen mit Dämonen'*. In A. Deuber-Mankowsky, U. Ramming, E. Walesca Tielsch (eds.). *1789/1989. Die Revolution hat nicht stattgefunden: Dokumentation des V. Symposions der Internationalen Assoziation von Philosophinnen*. Tübingen: Edition Diskord.

Lauterbach, H. (1991). „Trance, Magie und Anderswelterfahrung bei Bettine von Arnim und Annette von Droste-Hülshoff". *Schlangenbrut* 9, 15-24.

Lefebvre, H. (1987). *Kritik des Alltagslebens* (1946/1977). Frankfurt a. M.: Fischer.

Mader, E. (2002). *Die Macht des Jaguars. Natur und Weltbild der Shuar und Achuar in Amazonien*. In A. Gingrich et al. (eds.). *Metamorphosen der Natur. Sozialanthropologische Untersuchungen zum Verhältnis von Weltbild und natürlicher Umwelt*. Wien-Köln-Weimar: Böhlau, 183-222.

Majunke, C. (1999). *Der Genius loci: Geist des Ortes oder verorteter Geist. Landschaftsplanung zwischen dem Wunsch nach Ganzheit und moderner Subjektivität*. Berlin.

Mallien, L., Heimrath, J. (eds.). (2009). *Genius Loci. Der Geist von Orten und Landschaften in Geomantie und Architektur*. Klein Jasedow: Drachen.

Metzger, W. (1941). *Psychologie*. Dresden-Leipzig: Steinkopff.

Möckel, C. (2005): *Das Urphänomen des Lebens – Ernst Cassirers Lebensbegriff*, (Cassirer-Forschungen vol. 12). Hamburg: Meiner.

Nebel, G. (1983). „Heilung in Epidauros". *Merian*, 6.

Neddens, M. C. (1986). *Ökologisch orientierte Stadt- und Raumentwicklung: Genius loci, Leitbilder, Systemansatz, Planung*. Wiesbaden-Berlin: Bauverlag.

Norberg-Schulz, C. (1980). *Genius Loci. Towards a Phenomenology of Architecture*, New York: Rizzoli.

Norberg-Schulz, C. (1982). *Genius Loci. Landschaft, Lebensraum, Baukunst.* Stuttgart: Klett-Cotta.

Otto, W. F. (1910). *Genius*. In G. Wissowa, W. Kroll (eds.). *Paulys Real-Encyclopädie der classischen Altertumswissenschaft (Neue Bearbeitung)*, vol. 13. Stuttgart: Metzler, cols. 1155-1170.

Pedroli, B. (ed.). (2000). *Landscape – our Home / Lebensraum Landschaft. Essays on the Culture of the European Landscape as a Task / Essays über die Kultur der europäischen Landschaft als Aufgabe*, Stuttgart: Freies Geistesleben.

Pedone, S.,Tedeschini, M. (eds.). (2017). *Genius loci*. Milano: Mimesis (= Sensibilia 9/2017).

Pescarin, S. (1999): *Rom. Antike Bauwerke der ewigen Stadt*. Erlangen: Müller.

Pescarin, S. (2000): *Rome. Archaeological Guide to the Eternal City*. Milan: White Star.

Philippson, P. (1939). *Griechische Gottheiten in ihren Landschaften*. Oslo: Brøgger.

Pieper, J. (ed.). (1984). *Ort, Erinnerung, Architektur. Über den Genius Loci.* „Kunstforum" 69.

Pogačnik, M. (1995). *Elementarwesen. Die Gefühlsebene der Erde*. München: Knaur.

Preller, L. (1997). *Römische Mythologie* (1858). Essen: Phaidon. Reprographic reprint of the first edition.

Preußer, H.-P. (2015). *Pathische Ästhetik. Ludwig Klages und die Urgeschichte der Postmoderne.* Heidelberg: Winter.

Radke, G. (1979). *Tifata*. In K. Ziegler, W. Sontheimer (eds.). *Der kleine Pauly. Lexikon der Antike*, vol. 5. München: dtv, cols. 823f.

Rauh, A. (2012). *Die besondere Atmosphäre. Ästhetische Feldforschungen.* Bielefeld: Transcript.

Rauh, A. (2018). *Concerning Asthonishing Atmospheres* (Atmospheric Spaces 4). Milan, Mimesis.

Reisinger, C. (1991). *Die Hafeninsel in Saarbrücken*. In Die Gartenkunst No. 1, 73-101.

Relph, E. (1976). *Place and Placelessness*, London: Pion.

Relph, E. (2009). „A Pragmatic Sense of Place". *Environmental and Architectural Phenomenology*, 3, 24-31.

Rieche, A. (1978). *Das antike Italien aus der Luft*. Bergisch Gladbach: Deutsche Buch-Gemeinschaft.

Rohde, E. (1925). *Psyche. The Cult of Souls and Belief in Immortality among the Greeks.* Trans. by W. B. Hillis. London: Kegan etc. – German first edition: Rohde, E. (1890–1894). *Psyche. Seelencult und Unsterblichkeitsglaube der Griechen.* Freiburg: Mohr.

Röhnert, J. (2025). *Wildnisarbeit. Schreiben, Tun und Nature Writing*. Wuppertal: Arco.

Romeo, I. (1997). *Genius*. In *Lexicon Iconographicum Mythologiae Classicae* (LIMC), vol. 8 (suppl.). Zürich-Düsseldorf: Artemis.

Roscher, W. H. et al. (1886-1937) (eds.). *Ausführliches Lexikon der griechischen und römischen Mythologie*, 6 vols. Leipzig: Teubner.

Scheese, D. (1996). *Nature Writing. The Pastoral Impulse in America*. London-New York: Routledge.

Schilling, R. (1978) *Genius*. In T. Klauser (ed.). *Reallexikon für Antike und Christentum*, vol. 10. Stuttgart: Hiersemann, cols. 67-83.

Schlich, J. (2003). „Der ‚Geist der Heilung'. Bettina von Arnims homöopathische Politik in Die Günderode (1840)". *Internationales Jahrbuch der Bettina-von-Arnim-Gesellschaft* 15, 67-87.

Schmitz, H. (1965). *System der Philosophie. Vol. 2, 1: Der Leib*. Bonn: Bouvier.

Schmitz, H. (1967). *System der Philosophie. Vol. 3, 1: Der leibliche Raum*. Bonn: Bouvier.

Schmitz, H. (1969). *System der Philosophie. Vol. 3, 2: Der Gefühlsraum*. Bonn: Bouvier.

Schmitz, H. (1977). *System der Philosophie. Vol. 3, 4: Das Göttliche und der Raum*. Bonn: Bouvier.

Schmitz, H. (1978). *System der Philosophie, Bd. 3, 5: Die Wahrnehmung*. Bonn: Bouvier.

Schmitz, H. (1992). *Leib und Gefühl* (1989). Paderborn: Junfermann.

Schmitz, H. (1995). *Der unerschöpfliche Gegenstand. Grundzüge der Philosophie* (1990). Bonn: Bouvier.

Schmitz, H. (1998). *Der Leib, der Raum und die Gefühle*. Ostfildern: Edition Tertium.

Schmitz, H. (2002). *Hermann Schmitz, the "New Phenomenology"*. In A.-T. Tymieniecka (ed.). *Phenomenology World-Wide*, Boston-London: Kluwer, 491-494.

Schmitz, H. (2003). *Was ist Neue Phänomenologie?* Rostock: Koch 2003.

Schmitz, H. (2019). *New Phenomenology. A Brief Introduction* (Atmospheric Spaces 6). Milan: Mimesis.

Schmitz, H. (2023). *Atmospheres* (Atmospheric Spaces 13). Milan, Mimesis.

Scott, E. (ed.). (1890). *Calpurnius Siculus. The Eclogues*. London: George Bell and Sons.

Scully, V. (1962). *The Earth, the Temple and the Gods. Greek Sacred Architecture*. New Haven-London: Yale Univ. Press.

Seamon, D. (1980). *Body-subject, Time-space Routines, and Place-ballets*. In A. Buttimer, D. Seamon (eds.). *The Human Experience of Space and* Place. New York: Routledge, 148-165.

Seamon, D (2018). *Life Takes Place. Phenomenology, Lifeworlds, and Place Making*. London-New York: Routledge.

Seamon, D. (2020). "Place". *International Lexicon of Aesthetics*. https://lexicon.mimesisjournals.com/international_lexicon_of_aesthetics_item_detail.php?item_id=90 (accessed July 25, 2025).

Seamon, D. (2022). *Sense of Place*. In D. Richardson et al. (eds.). *The International Encyclopedia of Geography*. Hoboken: John Wiley & Sons.

Seamon, D (2023). *Phenomenological Perspectives on Place, Lifeworlds, and Lived Emplacement*. London-New York: Routledge.

Sloterdijk, P. (1998). *Sphären I – Blasen, Mikrosphärologie*. Frankfurt: Suhrkamp.

Sloterdijk, P. (2011). *Bubbles. Spheres vol 1: Microspherology.* Los Angeles: Semiotext(e), 2011.

Spitteler, G. (2014). "Dichte Teilnahme und darüber hinaus" (Thick Participation and Beyond), *Sociologus*, 64, 207-230.

Spitz, R. A. (1965). *The First Year of Life. A Psychoanalytic Study of Normal and Deviant Development of Object Relations*. New York: International Universities Press.

Spitz, R. A. (1972): *Vom Säugling zum Kleinkind. Naturgeschichte der Mutter-Kind-Beziehungen im ersten Lebensjahr.* Stuttgart: Klett.

Tedeschini, M. (2017). "La differenza del genio. Problema teorico, soluzione estetica". *Sensibilia*, 9, 351-365.

Thibaud, J.-P. (2022). "Ambiance". *International Lexicon of Aesthetics*, https://lexicon.mimesisjournals.com/international_lexicon_of_aesthetics_item_detail.php?item_id=113 (accessed July 5, 2025).

Thomson, H. J. (1953) (ed.). *Prudentius with an English Translation,* vol. 2. London: Heinemann.

Tilley, C. (2004). *The Materialiy of Stone. Explorations in Landscape Phenomenology*. Oxford-New York: Berg.

Tuan, Y.-F. (1974). *Topophilia. A Study of Environmental Perception, Attitudes, and Values.* Englewood Cliffs: Prentice-Hall.

Tuan, Y.-F. (1976). *Geopiety. A Theme in Man's Attachment to Nature and to Place*. In D. Lowenthal, M. J. Bowden (eds.). *Geographies of the Mind. Essays in Historical Geosophy in Honour of John Kirtland Wright*. New York: Oxford University Press.

Türcke, C. (2006). *Heimat. Eine Rehabilitierung,* Springe: zu Klampen.

Tworuschka, U. (ed.). (1994). *Heilige Stätten*. Darmstadt: WBG 1994.

Tymieniecka, A.-T. (1988). *Logos and Life, vol 1: The Creative Experience and the Critique of Reason.* Dordrecht-Boston-London: Springer.

Tymieniecka, A.-T. (1990). "The Human Condition within the Unity-of-Everything-there-is-Alive". *Analecta Husserliana*, 31, 3-17.

Valena, T. (2009). *Zu einer Phänomenologie des Genius Loci*. In L. Mallien, J. Heimrath (eds.). *Genius Loci. Der Geist von Orten und Landschaften in Geomantie und Architektur*. Klein Jasedow: Drachen, 148-206.

Walter, H. (1990). *Das griechische Heiligtum, dargestellt am Heraion von Samos*, Stuttgart: Urachhaus.

Watsuji, T. (1997). *Fudo – Wind und Erde. Der Zusammenhang von Klima und Kultur* (1935). Darmstadt: WBG.

Wrede, A. (2000): *Maria*. In H. Bächtold-Stäubli (ed.). *Handwörterbuch des deutschen Aberglaubens* (1935), vol 5. Berlin-New York: de Gruyter, col. 1649.

Ziegler, K. (1979). *Mundus*. In K. Ziegler, W. Sontheimer (eds.). *Der kleine Pauly. Lexikon der Antike*, vol 3. München: dtv, cols. 1463f.

MIMESIS GROUP
www.mimesis-group.com

MIMESIS INTERNATIONAL
www.mimesisinternational.com
info@mimesisinternational.com

MIMESIS EDIZIONI
www.mimesisedizioni.it
mimesis@mimesisedizioni.it

ÉDITIONS MIMÉSIS
www.editionsmimesis.fr
info@editionsmimesis.fr

MIMESIS COMMUNICATION
www.mim-c.net

MIMESIS EU
www.mim-eu.com

www.ingramcontent.com/pod-product-compliance
Lightning Source LLC
Chambersburg PA
CBHW031349230926
51848CB00002B/454
* 9 7 8 8 8 6 9 7 7 5 1 1 6 *